Origins of Human Innovation and Creativity

Developments in Quaternary Science
Series editor: Jaap J.M. van der Meer

Volumes in this series

1. The Quaternary Period in the United States
   Edited by A.R Gillespie, S.C. Porter, B.F. Atwater
   0-444-51470-8 (hardbound); 0-444-51471-6 (paperback) – 2004

2. Quaternary Glaciations – Extent and Chronology
   Edited by J. Ehlers, P.L. Gibbard
   Part I: Europe ISBN 0-444-51462-7 (hardbound) – 2004
   Part II: North America ISBN 0-444-51592-5 (hardbound) – 2004
   Part III: South America, Asia, Australasia, Antarctica ISBN 0-444-51593-3 (hardbound) – 2004

3. Ice Age Southern Andes – A Chronicle of Paleoecological Events
   By C.J. Heusser
   0-444-51478-3 (hardbound) – 2003

4. Spitsbergen Push Moraines – Including a translation of K. Gripp: Glaciologische und geologische Ergebnisse der Hamburgischen Spitzbergen-Expedition 1927
   Edited by J.J.M. van der Meer
   0-444-51544-5 (hardbound) – 2004

5. Iceland – Modern Processes and Past Environments
   Edited by C. Caseldine, A. Russell, J. Hardardóttir, Ó. Knudsen
   0-444-50652-7 (hardbound) – 2005

6. Glaciotectonism
   By J.S. Aber, A. Ber
   978-0-444-52943-5 (hardbound) – 2007

7. The Climate of Past Interglacials
   Edited by F. Sirocko, M. Claussen, M.F. Sánchez Goñi, T. Litt
   978-0-444-52955-8 (hardbound) – 2007

8. Juneau Icefield Research Project (1949–1958) – A Retrospective
   By C.J. Heusser {
   978-0-444-52951-0 (hardbound) – 2007

9. Late Quaternary Climate Change and Human Adaptation in Arid China
   Edited by David B. Madsen, Chen Fa-Hu, Gao Xing
   978-0-444-52962-6 (hardbound) – 2007

10. Tropical and Sub-Tropical West Africa – Marine and Continental Changes During the Late Quaternary
    By P. Giresse
    978-0-444-52984-8 – 2008

11. The Late Cenozoic of Patagonia and Tierra del Fuego
    Edited by J. Rabassa
    978-0-444-52954-1 – 2008

12. Advances in Quaternary Entomology
    By S.A. Elias
    978-0-444-53424-8 – 2010

13. The Mýrdalsjökull Ice Cap, Iceland. Glacial Processes, Sediments and Landforms on an Active Volcano
    Edited by A. Schomacker, J. Krüger, K.H. Kjær
    978-0-444-53045-5 – 2010

14. The Ancient Human Occupation of Britain
    Edited by Nick Ashton, Simon Lewis, Chris Stringer
    978-0-444-53597-9 – 2010

15. Quaternary Glaciations – Extent and Chronology. A Closer Look
    Edited by Jürgen Ehlers, Philip L. Gibbard, Philip D. Hughes
    978-0-444-53447-7 – 2011

16. Origins of Human Innovation and Creativity
    Edited by Scott Elias
    978-0-444-53821-5 – 2012

For further information as well as other related products, please visit the Elsevier homepage (http://www.elsevier.com)

Developments in Quaternary Science, 16

Series editor: Jaap J.M. van der Meer

Origins of Human Innovation and Creativity

Edited by

Scott Elias
Department of Geography
Royal Holloway, University of London,
United Kingdom

AMSTERDAM • BOSTON • HEIDELBERG • LONDON • NEW YORK • OXFORD
PARIS • SAN DIEGO • SAN FRANCISCO • SINGAPORE • SYDNEY • TOKYO

ELSEVIER

GUELPH HUMBER LIBRARY
205 Humber College Blvd
Toronto, ON   M9W 5L7

Elsevier
Radarweg 29, PO Box 211, 1000 AE Amsterdam, The Netherlands
The Boulevard, Langford Lane, Kidlington, Oxford, OX51GB, UK

Copyright © 2012 Elsevier B.V. All Rights Reserved.

No part of this publication may be reproduced, stored in a retrieval system, or transmitted in any form or by any means, electronic, mechanical, photocopying, recording, or otherwise, without the prior written permission of the publisher.

Permissions may be sought directly from Elsevier's Science & Technology Rights Department in Oxford, UK: phone (+44) (0) 1865 843830; fax (+44) (0) 1865 853333; email: permissions@elsevier.com. Alternatively you can submit your request online by visiting the Elsevier web site at http://elsevier.com/locate/permissions, and selecting Obtaining permission to use Elsevier material.

Notice
No responsibility is assumed by the publisher for any injury and/or damage to persons or property as a matter of products liability, negligence or otherwise, or from any use or operation of any methods, products, instructions or ideas contained in the material herein.

Library of Congress Cataloging-in-Publication Data
A catalog record for this book is available from the Library of Congress

British Library Cataloguing-in-Publication Data
A catalogue record for this book is available from the British Library.

ISBN: 978-0-444-53821-5
ISSN: 1571-0866

For information on all Elsevier publications
visit our web site at store.elsevier.com

Printed and bound in Great Britain

11 12 13   10 9 8 7 6 5 4 3 2 1

Working together to grow
libraries in developing countries

www.elsevier.com | www.bookaid.org | www.sabre.org

ELSEVIER   BOOK AID International   Sabre Foundation

Contents

Contributors vii

1. Origins of Human Innovation and Creativity: Breaking Old Paradigms   1
   Scott Elias

2. Creativity and Complex Society Before the Upper Palaeolithic Transition   15
   Clive Gamble

3. North African Origins of Symbolically Mediated Behaviour and the Aterian   23
   Nick Barton and Francesco d'Errico

4. Personal Ornaments and Symbolism Among the Neanderthals   35
   João Zilhão

5. Invention, Reinvention and Innovation: The Makings of Oldowan Lithic Technology   51
   Erella Hovers

6. Emergent Patterns of Creativity and Innovation in Early Technologies   69
   Steven L. Kuhn

7. The Evolutionary Ecology of Creativity   89
   John F. Hoffecker

8. Climate, Creativity and Competition: Evaluating the Neanderthal 'glass ceiling.'   103
   William Davies

Index   129

Contributors

Numbers in paraentheses indicate the pages on which the authors' contrbutions begin.

Nick Barton *(23), Institute of Archaeology, University of Oxford, 36 Beaumont Street, Oxford OX1 2PG, UK*

William Davies *(103), Centre for the Archaeology of Human Origins, University of Southampton, Avenue Campus, Southampton, SO17 1BF, UK., swgd@soton.ac.uk*

Francesco D'Errico *(23), CNRS UMR 5199 PACEA, Université Bordeaux 1, avenue des Faculte's, F-33405 Talence, France; Department of Archeology, History, Cultural Studies and Religion, University of Bergen, Norway*

Scott Elias *(1), Geography Department, Royal Holloway, University of London*

Clive Gamble *(15), Faculty of Humanities (Archaeology), Building 65A, Avenue Campus, University of Southampton, Southampton SO17 1BF, Clive.Gamble@soton.ac.uk*

John Hoffecker *(89), Institute of Arctic and Alpine Research, University of Colorado, Boulder, CO 80309 USA*

Erella Hovers *(51), Institute of Archaeology, The Hebrew University of Jerusalem, Mt. Scopus Jerusalem 91905, Israel, hovers@mscc.huji.ac.il*

Steven Kuhn *(69), School of Anthropology, University of Arizona, Tucson, AZ 85721-0030 USA, skuhn@email.arizona.edu*

João Zilhão *(35), ICREA Research Professor at the University of Barcelona, (Seminari d'Estudis i Recerques Prehistòriques; Departament de Prehistòria, Història Antiga i Arqueologia; Facultat de Geografia i Història; C/Montalegre 6; 08001 Barcelona; Spain), joao.zilhao@ub.edu*

Chapter 1

Origins of Human Innovation and Creativity: Breaking Old Paradigms

Scott Elias,
Geography Department, Royal Holloway, University of London

Where does human innovation and creativity come from? How did it arise? Did it need a set of triggers, and, if so, what were they? Can we discern patterns in the creative thought process just by examining the artefacts (mostly stone tools) preserved in archaeological sites, or should we be using other methods to reconstruct this fascinating aspect of human history? These questions, and others, were addressed by a group of archaeologists at a symposium sponsored by the British Academy in September 2009. The symposium was so interesting, and the participating speakers were so stimulated by the topic, that we decided to develop the theme into this edited volume of papers. The chapters in this book are wide ranging, and approach these questions from many different angles, focussing on a variety of human species, study regions and time intervals (Fig. 1.1). Their papers certainly challenge, if not break, some old paradigms.

One of the ways to look at the origins of creativity and innovation is to examine the physical evolution of the human brain. This has been facilitated in recent years through the development of rather sophisticated 3-dimensional modelling of the size and shape of human brains, ranging from modern humans back through most of the ancestral species. If a fossil skull is available for a species, then the brain lodged in that skull can be reconstructed with surprising precision, so that the size and shape of the various lobes of the brain can be measured accurately. It is clear that brain size has increased throughout the course of human evolution. As is well known, humans have an exceptionally large brain relative to their body size. For example, the brain weight of humans is 250% greater than that of chimpanzees, while the human body is only 20% heavier. Our ancestors living 2–2.5 million years ago had an average brain weight of 400–450 g, while our more immediate ancestors living 200,000–400,000 years ago had an average brain weight of 1350–1450 g. This threefold increase in size represents one of the most rapid morphological changes in evolution, even though its genetic basis remains elusive (Zhang, 2003). It is generally believed that the evolution of larger brain size set the stage for the emergence of human language and other high-order cognitive functions, and that it was driven by adaptive selection (Decan, 1992), but, as Schoenemann (2006) noted, it is clear that the human brain is not simply a larger version of the brains of our primate relatives. Rather, there are disproportionate increases in some parts of the brain, such as the frontal lobe. The changing shape of the human brain should provide clues about the behavioural evolution of our species. The evolutionary costs of growing and maintaining these masses of neural tissue must have been offset by some sort of adaptive (reproductive) advantages to successive populations of ancestral humans. Surely one of the main advantages must have been an increased capacity for innovative and creative thought.

It turns out that the psychologists and physical anthropologists who study such things do not all agree on which parts of the brain are the source of creative thought. Some of those who study the functioning of the modern human brain assert that the centre of creativity is found in the frontal lobes of our species. If so, then the so-called "executive functions" of the frontal lobes may have facilitated the evolutionary ascendency of humans (Coolidge and Wynn, 2001). The mental activity of the frontal lobes is considered by some to give rise to "all socially useful, personally enhancing, constructive and creative abilities" (Lezak, 1982).

Other authors have pointed to other parts of the brain as playing a central role in the evolution of human cognition and innovative thought processes. For instance, a study of changes in the shape and size of the various parts of the brain in the genus *Homo* reached the conclusion that the development of the parietal lobes in modern humans is the only nonallometric difference between *Homo sapiens* and nonmodern taxa, and that this morphological change may have represented a discrete cognitive shift (Bruner, 2004). According to this study, the parietal cortex may have

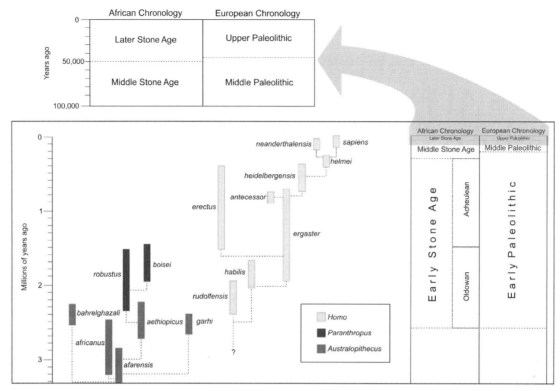

FIGURE 1.1 Timescale of adaptive radiations in humans (left, after Foley, 2002) and major divisions of the Pleistocene (right and above) in African and European archaeology.

played a principal role during hominin evolution, through its direct relationship with visuospatial integration, sensory integration, multimodal processing and social communication. A comparison of distances between noted landmark features of the brain showed changes in the parietal chord and the frontal chord through time. The ratio of these changes is shown in Fig. 1.2. This plot of *Homo erectus*, Neanderthal and anatomically modern human (AMH) brain features show an increase in the frontal/parietal chord ratio from *H. erectus* to *Homo neanderthalensis*, but the AMH specimens show increases in both chords, so the frontal/parietal chord ratio is less than that for *H. neanderthalensis*.

A review of human brain evolution studies by Schoenemann (2006) concluded that,

"Apart from cranial capacity, only suggestive, equivocal clues of possible behavioral patterns are evident in the fossil record of hominin brain evolution, mostly relating to the question of language evolution. Although definitive statements are not currently warranted, we do not presently know the limits of possible inferences about the behavior of fossil hominins from their endocranial remains."

Schoenemann (2006) suggested that the cognitive demands of tool making might have spurred brain evolution. He noted Reader and Laland's (2002) study showing that the frequency of tool use in primates is positively correlated with both absolute and relative brain volume. However, he also noted that research on the importance of stone-tool manufacturing in shaping the evolution of the

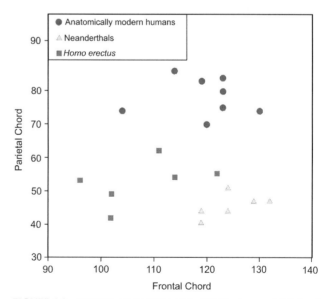

FIGURE 1.2 Bivariate comparisons of frontal chord vs. parietal chord distances, based on cranial measurements from fossil specimens of *Homo erectus*, *Homo neanderthalensis* and *Homo sapiens* (the latter including modern specimens). *After Bruner, 2004.*

human brain is in its infancy, and that future functional imaging studies are needed to clarify the issue.

Dunbar and Shultz (2007) considered the problem of primate brain evolution from an ecological standpoint, using the social brain hypothesis put forward by Byrne and Whiten (1988) that large brains accompany vertebrates with complex social lives. Viewing brain evolution from a broader ecological perspective, their view is that ecological problem solving (i.e., foraging strategies, offspring rearing and survival) are more effectively solved socially than individually in species exhibiting high degrees of sociality. Again, there is an ecological cost to maintaining a large brain, and this will only be worthwhile (and hence, the large brain trait passed on to succeeding generations) if the benefits outweigh the costs. By comparing the behavioural, ecological and life-history characteristics with the relative brain size of primates, they showed that brain volume does correlate strongly with sociality, and that the neocortex, the part of the brain made up of grey matter and divided into the frontal, parietal, occipital and temporal lobes, plays a critical role in the distinction between humans and other primates.

1.1. THE PROBLEM OF STASIS IN STONE TOOL TECHNOLOGY

The physical evidence of modern and ancestral human's brain size and shape has shed some light on the importance of this great mass of neurons in human evolution. But how do we get at the actual thought processes of our ancestors? These are not etched on the inside of a fossil skullcap. One of the great unsolved mysteries in palaeoanthropology/archaeology is the problem of stasis in stone-tool technology during the Early Pleistocene. After the initial invention of simple stone tools, seemingly little progress was made for an incredibly long time. Clive Gamble's chapter in this volume addresses this problem. Traditionally, archaeologists have attempted to solve such problems through detailed analysis of stone tools. But Gamble's thesis is that this problem cannot be addressed solely on the basis of the record of artefacts left behind by our ancestors. We must try to get into the minds of the early peoples and evaluate the role that their emotions played in driving innovation and creativity. When we focus just on the artefactual record, we are in danger of ignoring the basis of cultural interaction which depends on our sensory responses to both materials and people. Gamble argues that creativity is an embodied act and it is a social act. Early humans may have been enormously creative in their interactions with one another, even if their stone-tool kit failed to change for hundreds of generations. Of course, this places a new burden on archaeologists and palaeoanthropologists, as they attempt to deduce human behaviour and mentality from sources *other* than artefacts, but Gamble posits that this is not as impossible as it might seem. He poses the question of how the hominin mind should be modelled, and draws on Gosden's (2010) recent work, and the work of Grove and Coward (2008) and Coward and Gamble (2008), that our cognition is relational rather than rational; thus, hominin creativity does not appear just in symbolic items that were made and manipulated, but rather as a wider consideration of what being en-minded entails. In this context, the Palaeolithic mind is the same as the modern mind, or the mind of many social animals, in that it consists of body (senses and emotions) and things (both animate and inanimate) in reciprocal social partnerships rather than being directed solely by mental instructions sent from the brain. In other words, the brain, acting in concert with the rest of the body, and with tools or other material objects, forms a single unit: the mind in action.

Gamble then addresses the question of how many minds were needed for hominin creativity to evolve. Clearly, when it comes to the success and spread of innovations, the collective consciousness of a large group is more effective than that of the individual in any social species. This is one way in which being part of a larger community has improved the evolutionary fitness of humans throughout their history. We, and our ancestors, are, after all, rather puny, slow-moving, ineffectual organisms as individuals. Virtually all the predators and scavengers with which our ancestors competed on the ancient African savannas were faster, stronger and better equipped with large teeth and sharp claws. However, by working together in groups, and putting their minds together, our ancestors were able to invent survival strategies that led to their eventual climb to the top of the ecological pyramid. As discussed by Shennan (2001), when innovative minds feed off each other, the evolutionary fitness of the group improves.

How does the level of social interaction affect innovation and creativity? The size of an individual's active network apparently grew from about 80 individuals for *Australopithecus* to as many as 150 for our more recent ancestors. Gamble makes the point that larger groups require much higher levels of social interaction than small ones, and that one way humans have coped with this problem is to create hierarchies. Thus, an individual might have just five others in his or her support clique, 15 in his or her sympathy group, 50 in his or her band and 150 in his or her active network. Gamble et al. (2011) argue that this level of social complexity helped drive the expansion of human brain size.

Thus, Gamble argues that complex social behaviour must be acknowledged as a property of all the large-brained hominins that lived after 600,000 years ago, including *Homo heidelbergensis*, Neanderthals, late *H. erectus* and *H. sapiens*. Interestingly, Gamble suggests that what anthropologists call 'anatomically modern humans' might better be termed 'behaviourally modern humans'. The archaeology of the last 600,000 years indicates that human-made artefacts (*things*) did not get appreciably more complex until long after social complexity had occurred.

1.2. THE EVOLUTIONARY ECOLOGY OF CREATIVITY

Human beings, modern and ancient, have had the capacity to take existing elements of their environment and modify them to meet their needs. These essentially creative acts are, in some ways, what set us apart from other animals. We take a cobble and turn it into a chopper; we take a set of words and create new sentences from them. In John Hoffecker's chapter in this volume, he discusses the creative process in evolutionary and ecological contexts (the two have been tightly interwoven in the history of our ancestral species). Touching on a subject raised by Clive Gamble, Hoffecker stresses that hierarchical organisation has played a vital role in the development of the human mind. Our brains are wired to collect and then store vast quantities of information. Much of this is useless to us, but it gets stored, regardless of value. But we do make use of the *ability* to acquire and store information in our daily interactions with other people, and this social networking helped drive the evolution of increasing brain size, as discussed by Dunbar (1996).

The gathering and storage of information *per se* would not be worth nearly as much to humanity if we were not able to transform that information, to make it more useful. This is where creativity and innovation come in. Hoffecker notes that the most recent of the major breakthroughs in the evolution of life, as discussed by Maynard Smith and Szathmáry (1999), has been the development of human societies with language, out of primate societies that essentially lack language. Thus, neural information storage and retrieval, as expressed in human language, can be considered a major evolutionary breakthrough on this planet.

The honeybee is the only other species known to transmit complex hierarchically organised representations necessary for "information-centre" foraging. Their "language" is not verbal, but rather in the form of movements akin to human dance. The bee's communication is essentially a closed system, lacking the creativity that is found in human languages, with their infinite variety of word combinations.

As early humans spread across Africa and into other continents approximately 2 million years ago, they needed language to convey information to other members of their clan. Where can water be found? What game animals can be found here? Which plants are safe to eat, or effective to use as medicine? Hoffecker (2012) argues that the driving force behind the formation of the early human societies that dispersed out of Africa was likely the need to exchange information about such resources. This pattern of cooperative foraging is still seen among male–female pairs in modern human societies. A study by Hill (2002) examined the level of altruistic cooperation during food acquisition by the Ache people of Paraguay, one of the last surviving hunter–gatherer tribes. Her data showed that Ache men and women spend an average of about 10% of all foraging time engaged in altruistic cooperation; when pursuing game animals their cooperative foraging time rises to more than 50%. Hill concluded that cooperative food acquisition and redistribution in hunter–gatherer societies are critical behaviours that probably helped shape universal, evolved, cooperative tendencies.

Another breakthrough in the evolution of human processes is the ability to form a mental template – a means of ordering a series of actions to bring about the desired end. In order to make a hand axe, it is necessary to strike one or more blows in the right place. Not all archaeologists or palaeoanthropologists agree that the fashioning of such simple tools requires a mental template, but Hoffecker argues that by one million years ago, when biface tools were being made, such a template had to be functioning. The tool-manufacturing process involves three sequential, hierarchically organised steps, requiring a preconceived design. He goes on to argue that the essential difference between the minds of ancestral species and those of modern humans is that the latter exhibit a capacity for potentially unlimited recombination of informational units, hence creativity. By a quarter of a million years ago, humans were producing composite tools and weapons comprising three or four components (e.g., shafts, blades, binding cords and adhesive), with each component made from different raw materials and processed in a different way. Thus, tool making at this time involved the assembling of components in a hierarchically structured, preconceived design. But the designs were flexible, allowing new combinations of elements and materials. Hoffecker points to this phenomenon as a watershed in the history of human creativity, only to be superseded 200,000 years later by a new level of creativity expressed in the visual arts of the early Upper Palaeolithic in Europe.

1.2.1. Invention, Re-invention and Innovation: Makings of the Oldowan

In her chapter in this volume, Erella Hovers also attacks the problem of identifying creativity in the archaeological record of stone tools. She acknowledges that this problem is shared by all researchers of the Palaeolithic, stating that they have had to abandon attempts to pin down elusive bursts of creativity, those "firsts" that express creative potential, in the records they study. Rather, they have opted to look at the rates of turnover and degrees of variety in artefact forms. Her chapter focuses on the earliest of human technological innovations, the simple stone tools made by our ancestors 2.6–2.5 million years ago in East Africa, called Oldowan, after Olduvai Gorge, Tanzania, where they were first discovered by the Leakeys in the 1930s. As pointed out by Rogers and Semaw (2009), the Oldowan tools appear to be "something" that emerged out of "nothing". Thus, they mark the very beginnings of human creativity and innovation, as documented in the archaeological record. Oldowan technology represents the systematic flaking of stone, broadly associated with cutting activities. In ecological terms, stone-tool making was apparently a beneficial adaptation that helped shape the hominin niche for later species of *Australopithicus* and all species of the genus *Homo*. According to Hovers, this relatively simple invention spiralled into widespread innovations within a few hundred thousand years. Hovers addresses two questions in her chapter. First, does Oldowan technology represent a creative event? Are there elements of the Oldowan that may be legitimately considered as the outcome of creative acts? Second, was the spread of the Oldowan after 2.5 Ma due to social learning and cultural transmission, or was it independently reinvented by different groups, time after time, in separate creative acts?

The chapter begins with a discussion on *kinds* of creative thought. Hovers considers two fundamentally different modes, which might (to use modern vernacular phrases) be summarised as creative thinking outside the box, and creative thinking inside the box. The "box" in this case represents the social norms – the familiar thought patterns of a society, be it a small clan of a few individuals, or a larger group. She points out that these two kinds of creativity form the end members of a continuum. In terms of creativity in problem solving, thinking "inside the box" might represent attempts to come up with a solution, based on slight variations on existing methods. Thinking "outside the box" might represent attempts that are stabs in the dark. In human societies, such attempts are likely to be met with the sceptical comment, "But we've never done it that way, before." Armed with this conceptual framework, which I have attempted to summarise in but a few words, Hovers tries to assess whether it can be shown that Oldowan technology represents a real break from the behavioural repertoire that hominins share with apes (i.e., thinking "outside the box" in my summary), and thus an example of exceptional creativity, or whether this technology should be relegated to mundane creativity (thinking inside the box) if it is just an expansion of behaviours known among the apes.

The earliest known stone tools that have been deliberately flaked through percussive blows appear in localities in the Gona region of the Afar depression, northeast Ethiopia, at 2.6 Ma. By 2.4–2.3 Ma, the technology appears at sites in other regions of Ethiopia and Kenya. By 2.1–1.9 Ma, the sites have spread throughout the Rift Valley of East Africa and basins in North and Central Africa, and increased markedly in number. We cannot tell exactly who made these Oldowan tools, because at least four species of hominins (*Australopithecus garhi*, *Paranthropus boisei*, *Australopithecus aethiopicus* and *Homo* sp.) are known from East Africa between 2.5 and 2 million years ago, and between 2 and 1 million years ago, three more species of *Homo* appeared in the regional fossil record (*H. rudolfensis*, *H. habilis* and *H. erectus*). However, as discussed above, there is good physical evidence from brain case measurements that members of the genus *Homo* demonstrate a significant brain expansion during the period of 2.0–1.0 Ma. This may be linked with a tool making emphasis in their evolution and may have facilitated the ability of *H. erectus* to expand out of Africa by about 2 Ma.

While Oldovan stone-tool flaking technology changed little through this interval, the choice of raw materials from which the tools were made did change. People became more selective, over time. Hovers reviews the literature demonstrating that, in the early Oldowan, raw materials were chosen from local sources, and only transported small distances from their source. After about 2.0 Ma, raw materials were transported greater distances, and apparently were chosen more carefully. For instance, one notable change was a growing tendency to link certain raw materials to the production of certain kinds or shapes of tools. This shows more intentionality and greater knowledge, of both what kinds of stones work best in various tool types and where those kinds of stones can be found.

The validity of Oldowan assemblages older than about 2 Ma has been called into question because local cobbles can be broken into the same shapes as these simple tools through natural causes. Only when exotic-source stones are thus broken can we be more certain that a human agency is involved. Even when accumulations of broken animal bones occur in association with the older assemblages, this combination does not necessarily prove human agency, because the bones may have accumulated naturally along the water courses typically associated with Oldowan sites. Attempting to reconstruct two-million-year-old events is never easy, especially in dry environmental settings where only the rare wet places have much potential to preserve fossils. Stone-tool assemblages that traditionally have been considered to constitute archaeological "sites" in East Africa have thus come under special scrutiny. Hovers notes that the phenomenon of "sites" may be a by-product of the fact that hominins, unlike nonhuman primates, intensively engaged in activities that involved durable raw materials. These durable objects constitute virtually all we know of the material culture of early peoples. Even though it seems certain that these cultures included many perishable items, these have long since decayed, and are archaeologically silent.

Did early Oldowan toolmakers make use of their newly acquired technology to kill animals and process their meat and bone marrow? Hovers rejects this hypothesis, noting that there is little evidence for percussion, pitting or bone fracturing to access marrow in pre-2.0 Ma sites, and that securely identified cut marks associated with early Oldowan sites mainly come from surface finds, while isolated instances of cut marks from the Middle Awash region that are dated 2.5 Ma are not associated with lithic artefacts, making their interpretation difficult. It seems relatively clear that these early humans had meat in their diet, but how did they obtain it? In fact, they may either have been scavengers or have carved out a niche for themselves that lay somewhere between full-fledged predators and scavengers. Hovers notes the paucity of unambiguous evidence for meat consumption during the first 600,000 years of stone-tool making, and concludes that the makers of early Oldowan tools might just have used them to enhance their supply of plant foods.

Hovers compares and contrasts the tools made by other apes (e.g., chimpanzees) with those made by hominins. She concludes that the flaking of stone with the clear intention of creating a cutting-edged tool is uniquely human. She goes on to specify that in Oldowan stone knapping, familiar elements (stone, gestures of percussion and shapes of accidentally removed flakes with sharp edges) were linked in new ways, leading to useful solutions. Not only were cutting edges achieved, but they were also obtained in large numbers from a given core. Hovers concludes that the invention of Oldowan stone-tool making should be considered an act of exceptional creativity. *Australopithecus* thought outside the box. As Hovers more eloquently puts it, "The first Oldowan stone-tool making involved some breach of a cognitive 'glass ceiling'."

Finally, Hovers examines her second main question, whether the subsequent manifestations of Oldowan stone-tool technology represent reinvention or the transmission of knowledge through thousands of generations. While the *mode* of later invention remains obscure, Hovers concludes that the behaviours deduced from the younger Oldowan site assemblages represent expansions of previously known behaviours more than radically novel combinations of such behaviours. At best, these behaviours represent mundane creativity, or "thinking inside the box," as I put it, earlier.

1.3. EMERGENT PATTERNS OF CREATIVITY AND INNOVATION IN EARLY TECHNOLOGIES

As discussed above, increasing levels of human interaction are likely to have played a significant role in the development of human intellect, leading to increases in creative thought. The chapter by Steven Kuhn in this volume examines emergent patterns of creativity and increasing levels of interaction in early human societies, focussing on the demographics and network structures of those groups. He does not discount the creative spark of individuals, and its potential to cause great technological leaps forward; however, unfortunately, such individual acts of creation are essentially invisible in the archaeological record. Kuhn is interested in working out the origins of what he calls the "creative explosions" of the Eurasian Upper Palaeolithic and the later Middle Stone Age of Africa, to name just two. He asks, "What factors might affect the rate at which new things (artefacts, processes, etc.) make their appearance in the archaeological record?" He argues that this sort of emergent, aggregate creativity is a product of two broad sets of influences. The first is essentially biological. It is the cognitive capacity of hominins that has allowed them to produce new things and solve problems. But unless these bright ideas are communicated and become widely adopted, they are likely to perish before leaving their mark in the archaeological record. The second set of influences therefore involves the factors that foster the wider diffusion and persistence of such creative ideas.

How and when did major innovative breakthroughs occur? Kuhn warns us not to link such periods with the appearance of new hominin species. As discussed in Kuhn's chapter, as well as those of Hovers and Zilhão, trends in cultural evolution now appear to have been independent of these important biological transitions (see also Hovers and Kuhn, 2005). For instance, the Neanderthals of late Middle Palaeolithic Europe appear to have produced significant changes in material culture before the arrival of AMH. Kuhn also notes that not all technological advances persisted indefinitely in a given region, citing as an example the impressive cultural developments of the later Middle Stone Age in Southern Africa (ca. 77–58 ka) that were followed by an interval that retained few of these innovations and in fact resembles much older material cultures. In addition to different chronological trajectories for invention (great leaps forward or backward), Kuhn also discusses differences in geographic trajectories of invention, such as the different evolutionary trajectories seen in Lower and Middle Palaeolithic artefacts between East Asia and Western Europe.

Not all technological novelty is due to the same forces. Kuhn makes the distinction between random copying errors, a kind of genetic mutation in material culture, and purposeful, directed experimentation that brings about an innovation. The latter is a genuine form of creativity. Ultimately, it does not matter which kind of innovation is involved. If the result is sufficiently useful to a group of individuals, so much so that other groups take notice and want to copy it, then the invention has a good chance of persisting beyond the limited space and time of its creation. Transmission of innovative ideas takes place most easily in large groups of people, just as genetic mutations are more easily passed on to successive generations in large populations of a species. Kuhn points to the papers by Shennan

(2001) and Powell et al. (2009) that stress the link between demographic changes (i.e., times of substantial increases in human population in given regions) and times of major cultural or technological innovation. Conversely, times of population declines have been correlated with losses of elements of material culture in isolated populations. One of the overriding themes of Kuhn's chapter is, therefore, that progress in the invention of material culture has not been unidirectional. Rather, such advances can be either halted for long periods of time or even reversed during periods of population decline. He cites the earliest occupation of the southern fringes of Europe by people in the Lower Palaeolithic as an example. They left precious little evidence of technological innovation, but Kuhn argues that the uncertainties of living at the extreme edges of their species range could have made for small and unstable populations in which accumulated cultural knowledge could easily be lost by chance events. Thus, population stability may place as important a role in the pace of innovation as population size.

Networking is something at which modern humans are relatively good, and there is little doubt that some form of network building has been going on in humankind for hundreds of thousands of years. Of course, most human networks involve those closest to us, but long-distance networks can also be important in the transmission of new ideas, a fact emphasised by Kuhn as a means of buffering the negative influence of the dying out of small, local populations. Kuhn points out that hunter-gatherers maintain social networks with people outside their immediate clans, whether or not these outsiders are kin. Trade networks have a role to play here. The human desire to acquire exotic goods from distant regions has been demonstrated in the archaeological record, at least as far back as the Upper Palaeolithic (Barton et al., 1994). Kuhn identifies long-distance transport of highly prized workable stone, such as obsidian, in the late Middle Stone Age of Africa (roughly 75,000–50,000 yr ago).

Demographics play an increasingly important role in human cognition in the Upper Palaeolithic, according to Kuhn, because the level of social interactions rise, facilitating social learning. According to social learning theory (Bandura, 1977), humans learn from their environment. As long as humans were living in relatively small, isolated communities, social learning was likewise limited. But when populations rose, as they did at times in Upper Palaeolithic Europe, for instance, then the level of social interaction also rose, and with it came enhanced social learning. In this setting, small-scale innovations can accumulate and spread rapidly in a society.

Finally, Kuhn makes an interesting observation on the role of hardship in stimulating innovation. Again, linking this phenomenon with demographics, he observes that if environmental stress (e.g., the onset of a cold interval in Late Pleistocene Europe) is too great, rather than stimulating innovation, its main impact is to lower population size through death or migration. Smaller populations have fewer social interactions, which Kuhn considers an important driver in innovation and creativity.

1.3.1. Personal Ornaments and Symbolism Among the Neanderthals

Having considered the origins of creativity and innovation among the earlier species of hominins, we now shift our gaze to one of the most recent species, *H. neanderthalensis*, or Neanderthal man. This volume includes two chapters devoted to various aspects of Neanderthal creativity. The first of these is by João Zilhão. He takes the old paradigms about Neanderthal inventiveness to task. The discovery of the fossil skullcap and other bones of a non-AMH in the Neander valley of Germany in 1856 represents a watershed in palaeoanthropology. Victorian Europeans were not ready to consider that an extinct, ancient human might have been (a) closely related to modern humans and (b) our intellectual equals. One of the first artist's reconstructions of Neanderthal man, from 1909, is quite telling (Fig. 1.3). Here we see, literally, an ape-man, covered with fur, crouched like a gorilla, with the look of a dangerous wild animal in its eyes. Both the general public and, to a large extent, palaeoanthropologists of previous generations always emphasised the "other-ness" of Neanderthals. Until the discovery of a Neanderthal hyoid bone in Kebara Cave, Israel (Bar-Yosef et al., 1992), most palaeoanthropology textbooks stated that Neanderthals had little or no power of speech, much less anything resembling modern human language. Artefactual evidence contradicting this ape-man paradigm was either ignored by most researchers, or ascribed to modern humans, in spite of stratigraphic or dating evidence to the contrary. This line of thinking was decidedly circular: "Neanderthals were not intelligent or creative enough to have made this set of artefacts, therefore they *must* have been made by modern humans." Zilhão systematically challenges the old paradigms, taking a fresh look at the archaeological evidence.

In Zilhão's chapter, he identifies two statements concerning indicators of "behavioural modernity" in the African archaeological record whose validity is widely accepted by palaeoanthropologists. First is a definitive statement by Henshilwood and Marean (2003): "Artefacts or features carrying a clear, exosomatic symbolic message, such as personal ornaments, depictions, or even a tool clearly made to identify its maker." Second is a different take on modernity by Brown et al. (2009): "Complex use of technology, namely the controlled use of fire as an engineering tool to alter raw-materials; for example, heat pre-treating poor-quality siliceous rocks to enhance their flaking properties". Zilhão argues that, if these statements may be used

FIGURE 1.3 Artist's reconstruction of the Neanderthal 'Man of Chapelle-Aux-Saints' by Mr. Kupka (scientifically advised by Marcellin Boule) for the Illustrated London News, 27 February, 1909.

to define modernity, then any clear-cut evidence that Neanderthals fit these descriptions *must* be taken as evidence of their behavioural modernity.

The first instance of archaeologists and palaeoanthropologists ignoring or discounting evidence of the use of symbolism amongst Neanderthals comes from the French rockshelter of La Ferrassie, in the Dordogne (Peyrony, 1934; Defleur, 1993). Here, the remains of seven Neanderthals were found in a single level of a deeply stratified deposit. Based on cultural indicators linked with other sites in France, these people were buried between 60,000 and 75,000 years ago. These facts are undisputed, but what was overlooked for more than 70 years were two significant features of these burials: a bone fragment decorated with four sets of parallel incisions that lay next to an adult male, and a limestone slab decorated with cup holes that was buried in a deep pit with a child.

Zilhão also cites a prime example of the use of complex technology by Neanderthals, from the open-air site of Königsaue, eastern Germany (Mania, 2002). This site yielded fragments of birch bark pitch, one of which bore a human fingerprint as well as impressions of a flint blade and of wood-cell structures, indicating use as an adhesive material to fix a wooden haft to a stone knife. Radiocarbon dating yielded an age greater than 50,000 yr BP, and chemical analysis showed that the pitch had been produced through a lengthy smouldering process requiring a strict protocol: oxygen was excluded and the temperature had to be 340–400 °C to get this product. At the Italian site of Campitello, birch bark pitch remains date to >120,000 yr BP, making this pitch the first known artificial raw material in the history of humankind. Zilhão states that the sophistication of the fire technology employed in the production of this adhesive pitch remained unsurpassed until the invention of Neolithic pottery kilns. Zilhão argues that these and other examples should have led palaeoanthropologists to the conclusion that Neanderthals were more behaviourally advanced than AMH, but the dominant theme in the discussion remained the idea that Neanderthals went extinct in the Late Pleistocene, so they must have somehow been inferior to the AMH with which they cohabited Europe during the last glacial interval. The "out of Africa" and "human revolution" concepts drove palaeoanthropology, and the attention of researchers, away from the evidence discussed above.

One method of discounting Neanderthal use of symbolism is to argue that this was a very late development amongst Neanderthals, based on their imitation of behaviours exhibited by the AMH co-occupying Europe in the Late Pleistocene. A second method used to explain away Neanderthal use of symbolism is the two-pronged argument that (a) the sites where the artefacts might suggest Neanderthal use of symbolism are actually very rare and (b) in these rare cases, overlying AMH artefacts have been reworked to lower (Neanderthal) occupation layers.

Zilhão argues that the first method of explaining away Neanderthal use of symbolism (imitation of objects and behaviours of AMH cohabiting Europe) can now be firmly put to rest on the basis of improved chronostratigraphic control of European Upper Palaeolithic sites, coupled with improved radiometric dating of sites. These refinements show that the emergence of the Châtelperronian (Neanderthal) culture predates the earliest Aurignacian (AMH) culture as well as the earliest skeletal evidence for AMH throughout Europe by many thousands of years. He attacks the second argument (intrusion of AMH symbolic artefacts into Neanderthal layers at multioccupation sites) by positing that there are many sites where symbolic objects are clearly associated with Neanderthal occupation; so, even if attempts to discredit single sites are successful, these do not negate the whole body of evidence. He also cites Caron et al. (2011), in a detailed argument refuting attempts to relegate Neanderthal symbolic artefacts at the Grotte du Renne site in France as intrusive elements from overlying AMH layers, stating that this is completely inconsistent with their vertical distribution across the site's stratigraphic sequence. He cites supporting evidence from other Châtelperronian sites in Europe, including Quinçay rock shelter in France, the Ilsenhöhle rock shelter in Germany, the Trou Magrite (Pont-à-Lesse) site in Belgium, Bacho Kiro cave (Dryanovo) in Bulgaria, Klisoura 1 cave in Prosymna, Greece and Fumane rockshelter in Molina, Italy. In each case, artefacts with clearly symbolic significance have been found in Châtelperronian contexts, and each assemblage has been dated from 41,000 to 45,000 calendar years ago.

Zilhão et al., 2010 have also described Mousterian-level perforated shells from two Spanish caves (Cueva de los Aviones and Cueva Antón). These assemblages date back to 50,000 yr BP. One oyster shell had been used as a container for the storage or preparation of a complex cosmetic pigment most likely used as body paint. There were lumps of iron pigments of different mineral species (hematite, goethite and siderite), and especially yellow natrojarosite, whose only known use is in cosmetics. They also found an unmodified pointed bone bearing pigment residues on the broken tip, suggesting its use in the preparation or application of colourants. Zilhão argues in this volume that these artefacts, plus similar finds of pigments from Pech de l'Azé, Carsac-Aillac, France and the middle Palaeolithic in the Qazfeh Cave, Israel, present clear evidence for the use of body paint by Neanderthals.

In recent years, the evidence discussed above has been more widely accepted by both the general public and by the scientific community. Zilhão suggests that this new acceptance corresponds to the publishing of the first results of the Neanderthal genome project, published at about the same time (Green et al., 2010). The ancient DNA evidence indicates that modern humans share some DNA with Neanderthals, indicating that there must have been some interbreeding between the two groups in the Late Pleistocene. This, in turn, removed the need for considering AMH and Neanderthals as completely separate, competing species. If the "other-ness" of Neanderthals, the prevailing paradigm for 150 years of palaeoanthropology, could be at least weakened, if not nullified by the DNA evidence, then this helped eliminate the barriers to acknowledging that Neanderthals had fully human cognition, as evidenced by artefacts indicating a fully symbolic material culture. Not all archaeologists or palaeoanthropologists are convinced, of course.

Old paradigms die hard, in archaeology and elsewhere. The earliest impressions of Neanderthals were that they were hairy ape-men. As Zilhão so aptly says in his chapter, "You never get a second chance to make a first impression."

1.3.2. Climate, Creativity and Competition: Evaluating the Neanderthal "Glass Ceiling"

Zilhão's chapter provides evidence that, at least in their last few millennia of existence, Neanderthals had fully human cognition and behaviour. In the second chapter concerning Neanderthals, William Davies approaches Neanderthal archaeology from a different viewpoint. He asks how they changed and developed over the course of their long (200,000-year) existence, how they innovated and interacted and how climate and competition affected them. Davies' chapter in this volume thus takes a longer view of *H. neanderthalensis*. Using an environmental approach, Davies places Neanderthal innovations in the context of environmental pressures. He proposes that these innovations were fairly constantly generated in the Palaeolithic, but selection pressures were more severe against them in times of environmental adversity. Thus, climate might have exerted little pressure on the *rate* of innovation, but rather more on the fate of those innovations.

Davies constructs a theoretical framework by which to judge Neanderthal innovation throughout the course of their history. He considers change at different material culture scales (from small to large), and in terms of its structure and social organisation. Small-scale changes should have the highest turnover (perhaps days or weeks) and be restricted to the intrasite scale, while large-scale change should occur over longer periods, ranging from decadal to centennial spread across wide areas.

Demography (e.g., population density and social structure) needs to be considered in the development and transmission of novel traits. Larger populations may increase the number and rate of invention, as discussed in relation to Kuhn's chapter, but Davies argues that social structure and connectedness are a better explanation for transmission of new ideas, especially in mobile populations. Mobility is another important factor. The effect of different scales of mobility needs to be recognised in the generation and transmission of novel traits. More sedentary populations will transmit (or not) their ideas in different ways, and perhaps at different rates, than more mobile populations. Innovations may spread through face-to-face contact or they can spread greater distances through social networks. Davies laments the low number of reliable chronologies for Neanderthal sites, especially for sites beyond the range of radiocarbon dating (i.e., 50,000 yr BP). Without accurate dates for sites and artefacts, the *pace* of innovation and transmission of ideas cannot be determined.

Davies defines various categories of Neanderthal innovations, so that each can be independently assessed. These include burial of the dead, the use of different parts of sites for different functions, blade and bladelet technology, the use of bone, antler and ivory artefacts and tools, the possible use and manufacture of "symbolic" artefacts, e.g., beads, and efficient, specialised hunting strategies. The evidence for Neanderthal burial of their dead mostly comes from southwest France, the Ardennes, the Crimea, the Levant and the Zagros Mountains. The majority of burials are dated between 70,000 and 40,000 BP. Other methods of disposal of corpses were also practiced, and Davies concludes that there is no clear-cut pattern of transmission of burial practice innovation, either spatially or temporally. There is no also clear-cut pattern of Neanderthal use of different parts of their habitation sites for different purposes. Davies concludes that this apparent lack of differentiation of living spaces means that there were no *specific* locations for transmission of novel ideas and techniques in Neanderthal sites, though perhaps such transmission might be concentrated in the areas of greatest activity. Neanderthal blade production demonstrates flexible, fluid knapping strategies, and their production of "nanopoint" bladelets less than 1 cm long demonstrates Neanderthal innovation, dexterity and technical precision. Davies argues that the use of ivory, antler, bone and shell artefacts was both spatially limited (mostly to northern France) and temporally limited (between about 45,000 and 36,500 BP). Further, he argues that the use of beads and pendants is not ubiquitous in Neanderthal sites. Finally, concerning specialised hunting strategies, Davies weighs the evidence and concludes that, while there is evidence of butchery of prime individuals, there is no evidence of a trend in Neanderthal selection of prime animals over less fit individuals. The existing examples are both spatially and temporally discontinuous. Thus, it is difficult to gauge the extent and scale of Neanderthal innovation in hunting specialisation.

Davies argues that this lack of evidence for the spread of innovation could just as easily be applied to *H. sapiens* populations prior to ca. 50,000 BP. Only as population densities rose (between 50,000 and 35,000 BP) did AMH begin to consistently transmit innovations. Prior to 50,000 yr BP, populations of both species were too patchy in most regions to facilitate the transmission of new ideas. As the range of Neanderthals contracted after 50,000 BP into parts of Europe, their population densities may have risen, facilitating the spread of ideas in the last remaining groups. Davies notes that after 50,000 BP, both Neanderthals and AMH show increasing evidence for symbolic activity, e.g., bead production. His chapter argues that such increases in symbolic activity can be attributed to increased social interaction, but in the case of the Neanderthals, this was perhaps linked with increasing stress on social systems. However, Davies warns that while it might seem that the intensity of Neanderthal innovation and inventiveness increased after about 50,000–45,000 years ago, we cannot be sure at present if this apparent shift is an artefact of our dating, or a behavioural reality. The limitations of the radiocarbon method may be imposing a false time barrier on our comprehension of Neanderthal behaviour, because it is much more difficult to date artefact assemblages that are beyond this 50,000-year boundary.

On the question of Neanderthal imitation of behaviours exhibited by the AMH co-occupying Europe in the Late Pleistocene, Davies urges caution. He says that we cannot really say if Neanderthals and *H. sapiens* influenced each other's innovations, or whether they developed independently, and that it is difficult in many situations to distinguish independent (re)invention of characteristics from inter-Neanderthal acculturation.

Why did the spread of novel ideas, at least before 50,000 yr BP, take such a long time in Neanderthal societies? Davies describes the Neanderthal social world as comprising many small-scale closed networks, with limited exchange of information and ideas. This is clearly not the ideal substrate for either the transmission or the long-term persistence of new ideas and techniques. This social structure seems to be the best explanation available for the pattern of change seen in Neanderthal assemblages.

1.3.3. North African Origins of Symbolically Mediated Behaviour and the Aterian

In the final chapter of this book, Nick Barton and Francesco d'Errico focus their attention on the nature and timing of a few key innovations in the cultural record of North Africa during the Middle Palaeolithic/Middle Stone Age (MP/MSA). In developing the history of human innovation and creativity, North Africa was traditionally considered by archaeologists to have been "cul de sac", overshadowed by the more prolific records of East and sub-Saharan Africa. However, more recent attention has focused on the early use of symbolism in the Aterian industry of the MP/MSA. This industry has been found at sites spanning much of the Sahara, from the Atlantic coast of Morocco to Egypt and the Sudan. The bulk of the evidence indicates that the skeletons found at some Aterian sites represent AMH remains. The sites date back as far as 80,000 yr BP. The most widely accepted definition of the Aterian is that of Tixier (1967), based on stone tools described from Oued Djebbana, Algeria. He described the Aterian as a Levallois industry with a laminar or blade-like debitage showing a high proportion of faceted butts. In this method of stone-tool production, a striking platform is formed at one end of a core, and then the edges are trimmed by flaking off pieces around the perimeter. This creates a domed shape on the side of the core, known as a tortoise core, as the tool at this stage resembles a tortoise shell. When the platform is struck, a flake is driven off from the core that has a distinctive plano-convex profile. All of its edges are sharpened, due to the earlier edge trimming. As described by Barton and d'Errico, the Aterian toolkit includes side-scrapers and points, with a predominance of end-scrapers. Another important element are pedunculate tools that have a tang at their proximal end (Fig. 1.1). There is often bifacial thinning of the tanged ends, facilitating their mounting on shafts or handles. One of the most striking features of Aterian lithic technology is the co-occurrence of these thinned pedunculate tools and bifacially flaked, foliate points. The points appear to have been made as projectile points, but some of them are so small that they seem more likely to have been used on arrows, rather than on spears.

Recent excitement concerning the Aterian in the archaeological community has come about because of newly revised age estimates for this industry. Barton and d'Errico discuss optically stimulated luminescence (OSL) ages; in addition to uranium-series dates and thermoluminescence (TL) ages from sites in Morocco, they place this technology back to MIS 6, with TL dates from 145,000–171,000 yr BP. Further west on the Atlantic coast in the Témara district of Rabat, the earliest Aterian industry at one site has been OSL dated to MIS 5e, ca. 114,000–105,000 yr BP (Barton et al., 2009), and at a nearby site the layers containing Aterian artefacts date from 100,000 to 121,000 yr BP (Schwenninger et al., 2010). The Aterian has mostly been dated from 60,000 to 80,000 further east on the North African coast in Libya and Tunisia, although sites in Egypt have yielded uranium-series ages of $126,000 \pm 4000$ yr BP and an electron spin resonance (ESR) minimum age of $96,240 \pm 2500$ yr BP; so a clear geographic pattern for the spread of this lithic industry has not yet appeared.

Barton and d'Errico also report on the first finds of red pigments in Aterian assemblages in Morocco, dated 111,000–105,000 yr BP, and on the use of red ochre at several Moroccan cave sites, that date from 83,000 to 82,000 yr BP. Personal adornment is a key symbolic behaviour, and this has been found in Aterian sites in the form of shell beads. Barton and d'Errico discuss perforated marine snail shells of the taxa *Nassarius gibbosulus*, *Nassarius circumcintus* and *Columbella*. These finds date from 83,000 to about 60,000 yr BP in Moroccan cave sites. Interestingly, none of the shell bead artefacts dates from the oldest Aterian occupation layers at these sites.

Assessing the Aterian technocomplex in light of the changing environments of North Africa from MIS 6-3, Barton and d'Errico note that it may have persisted for more than 70,000 years, successfully enduring major environmental changes, and extending across a territory of 1,000,000 km^2. Most of the recently dated Aterian sites are associated with the interval of variable climate in the earlier phases of MIS 5 (MIS 5e-c), and the beginning of MIS 5a, which is characterised by a gradual decrease of temperatures and precipitation. The Aterian may have persisted through the cool, dry conditions of MIS 4, until the onset of MIS 3. There were few changes in the Aterian toolkit throughout its 70,000-year history. Barton and d'Errico conclude that the features that made the Aterian so successful can at the same time also be perceived as symptomatic of its inherent limitations. Contemporaneous and younger cultures in Europe, the Middle East and southern Africa provide better examples of the kind of rapid cultural changes we associate with modernity. These other regions apparently experienced environmental changes such as sea-level fluctuations that led to human habitat expansions and contractions, and there were intervals of population growth, during which the pace of technological change is more likely to increase (see the discussion of this in the section on Kuhn's chapter, above). However, sea-level changes had little effect on the coasts of North Africa, and the initial phases of each interglacial opened pathways through the Sahara (expansion of lakes and rivers), which paradoxically reduced the isolation of this region.

1.4. CONCLUDING REMARKS

The history of human creativity and innovation is very challenging to reconstruct. The authors contributing to this volume have, themselves, demonstrated considerable creativity in their manifold approaches to this thorny topic. It is often difficult, if not impossible, to determine *who* made a given artefact (i.e., which species of human) and exactly *when* it was made. The questions of *how* and *where* are often somewhat easier to address, but "why" is often the most difficult of all these questions. The authors have tried their best to "get into" the minds of ancient peoples. While the results of these endeavours are necessarily speculative, they may serve to advance the sciences of palaeoanthropology and archaeology, not least by challenging old assumptions and breaking old paradigms. A recurring theme in this volume is that increased levels of human interaction appear to be a powerful driver of creative thinking, as expressed in technological innovation. It is a great frustration that all we have to go on when we set out to interpret an ancient culture is a set of stone tools and a few other durable artefacts. Not only do we know little or nothing about the perishable elements of these cultures (i.e., clothing and other textiles, nets, rope, etc.), but there are other vital aspects of their cultures we will never know about, unless a means of time travel is invented. What songs did they sing? What stories did they tell? How were children taught what they needed to know? But this is not to disparage the efforts to understand the origins of human creativity and innovation documented in this volume and elsewhere. These writings represent real progress in the difficult voyage of discovery.

REFERENCES

Bandura, A., 1977. Social Learning Theory. General Learning Press, New York.

Bar-Yosef, O., Vandermeersch, B., Arensburg, B., Belfer-Cohen, A., Goldberg, P., Laville, H., Meignen, L., Rak, Y., Speth, J.D., Tchernov, E., Tillier, A.-M., Weiner, S., 1992. The excavations in Kebara Cave, Mt. Carmel. Current Anthropology 33, 533–534.

Barton, C.M., Clark, G., Cohen, A., 1994. Art as information: explaining upper Paleolithic art in Western Europe. World Archaeology 26, 185–207.

Barton, R.N.E., Bouzouggar, A., Collcutt, S., Schwenninger, J.-L., Clark-Balzan, L., 2009. OSL dating of the Aterian levels at Dar es-Soltan I (Rabat, Morocco) and implications for the dispersal of modern *Homo sapiens*. Quaternary Science Reviews 28, 1914–1931.

Brown, K.S., Marean, C.W., Herries, A.I.R., Jacobs, R., Tribolo, C., Braun, D., Roberts, D.L., Meyer, M.C., Bernatchez, J., 2009. Fire as an engineering tool of early modern humans. Science 325, 859–862.

Bruner, E., 2004. Geometric morphometric and paleoneurology: brain shape evolution in the genus Homo. Journal of Human Evolution 47, 279–303.

Byrne, R.W., Whiten, A., 1988. Machiavellian Intelligence: Social Expertise and the Evolution of Intelligence in Monkeys, Apes and Humans. Oxford University Press, Oxford.

Caron, F., d'Errico, F., Del Moral, P., Santos, F., Zilhão, J., 2011. The reality of Neandertal symbolic behavior at the Grotte du Renne, Arcy-sur-Cure. PLoS ONE 6 (6), e21545. http://dx.doi.org/10.1371/journal.pone.0021545.

Coolidge, F.L., Wynn, T., 2001. Executive functions of the frontal lobes and the evolutionary ascendancy of *Homo sapiens*. Cambridge Archaeological Journal 11, 255–260.

Coward, F., Gamble, C., 2008. Big brains, small worlds: Material culture and the evolution of mind. Philosophical Transactions of the Royal Society B 363, 1969–1979.

Decan, T.W., 1992. Biological aspects of language. In: Jones, S., Martin, R., Pilbeam, D. (Eds.), The Cambridge Encyclopedia of Human Evolution. Cambridge University Press, Cambridge, pp. 128–133.

Defleur, A., 1993. Les Sépultures Moustériennes. CNRS, Paris.

Dunbar, R.I.M., 1996. Grooming, Gossip, and the Evolution of Language. Harvard University Press, Cambridge, Massachusetts.

Dunbar, R.I.M., Shultz, S., 2007. Understanding primate brain evolution. Philosophical Transactions of the Royal Society of London Series B 362, 649–658.

Foley, R.A., 2002. Adaptive radiations and dispersals in hominin evolutionary ecology. Evolutionary Anthropology 11, 32–37.

Gamble, C.S., Gowlett, J.A.J., Dunbar, R., 2011. The social brain and the shape of the Palaeolithic. Cambridge Archaeological Journal 21, 115–136.

Gosden, C., 2010. The death of the mind. In: Malafouris, L., Renfrew, C. (Eds.), The Cognitive Life of Things: Recasting the Boundaries of the Mind. McDonald Institute for Archaeological Research, Cambridge, pp. 39–46.

Green, R.E., Krause, J., Briggs, A.W., Maricic, T., Stenzel, U., Kircher, M., Patterson, N., Li, H., Zhai, W., Fritz, M.H.-Y., Hansen, N.F., Durand, E.Y., Malaspinas, A.-S., Jensen, J., Marques-Bonet, T., Alkan, C., Prüfer, K., Meyer, M., Burbano, H.A., Good, J.M., Schultz, R., Aximu-Petri, A., Butthof, A., Höber, B., Höffner, B., Siegemund, M., Weihmann, A., Nusbaum, C., Lander, E.S., Russ, C., Novod, N., Affourtit, J., Egholm, M., Verna, C., Rudan, P., Brajkovic, D., Kucan, Ž., Gušic, I., Doronichev, V.B., Golovanova, L.V., Lalueza-Fox, C., de la Rasilla, M., Fortea, J., Rosas, A., Schmitz, R.W., Johnson, L.F., Eichler, E.E., Falush, D., Birney, E., Mullikin, J.C., Slatkin, M., Nielsen, R., Kelso, J., Lachmann, M., Reich, D., Pääbo, S., 2010. A draft sequence of the Neandertal genome. Science 328, 710–722.

Grove, M., Coward, F., 2008. From individual neurons to social brains. Cambridge Archaeological Journal 18, 387–400.

Henshilwood, C., Marean, C., 2003. The origin of modern human behavior. Critique of the models and their test implications. Current Anthropology 44, 627–651.

Hill, K., 2002. Altruistic cooperation during foraging by the Ache, and the evolved human predisposition to cooperate. Human Nature 13, 105–128.

Hoffecker, J.F., 2012. The information animal and the super-brain. Journal of Archaeological Method and Theory published online December 2011.

Hovers, E., Kuhn, S. (Eds.), 2005. Transitions before the Transition: Evolution and Stability in the Middle Paleolithic and Middle Stone Age. Springer, New York.

Lezak, M.D., 1982. The problem of assessing executive functions. International Journal of Psychology 17, 281–297.

Mania, D., 2002. Der Mittelpaläolithische Lagerplatz am Aschersebener See bei Königsaue (Nordharzvorland). Praehistoria Thuringica 8, 16–75.

Maynard Smith, J., Szathmáry, E., 1999. The Origins of Life: From the Birth of Life to the Origin of Language. Oxford University Press, Oxford.

Peyrony, D., 1934. La Ferrassie. Moustérien – Périgordien – Aurignacien. Préhistoire 3, 1–92.

Powell, A., Shennan, S., Thomas, M., 2009. Late Pleistocene demography and the appearance of modern human behavior. Science 324, 1298–1301.

Reader, S.M., Laland, K.N., 2002. Social intelligence, innovation, and enhanced brain size in primates. Proceedings of the National Academy of Sciences USA 99, 4436–4441.

Rogers, M.J., Semaw, S., 2009. From nothing to something: the appearance and context of the earliest archaeological record. In: Camps, M., Chauhan, P.R. (Eds.), Sourcebook of Paleolithic Transitions. Methods, Theories, and Interpretations. Springer, New York, pp. 155–171.

Schoenemann, P.T., 2006. Evolution of the size and functional areas of the human brain. Annual Review of Anthropology 35, 379–406.

Schwenninger, J.-L., Collcutt, S.N., Barton, N., Bouzouggar, A., Clark-Balzan, L., El Hajraoui, M.A., Nespoulet, R., Debénath, A., 2010. A new luminescence chronology for Aterian cave sites on the Atlantic coast of Morocco. In: Garcea, E.A.A. (Ed.), South-eastern Mediterranean Peoples between 130,000 and 10,000 Years Ago. Oxbow Books, Oxford, pp. 18–36.

Shennan, S., 2001. Demography and cultural innovation: a model and its implications for the emergence of modern human culture. Cambridge Archaeological Journal 11, 5–16.

Tixier, J., 1967. Procédés d'analyse et questions de terminologie concernant l'étude des ensembles industriels du Paléolithique récent et de l'Epipaléolithique dans l'Afrique du nord-ouest. In: Bishop, W.W., Clark, J.D. (Eds.), Background to evolution in Africa. University of Chicago Press, Chicago, pp. 771–820.

Zhang, J., 2003. Evolution of human APSM gene, a major determinant of brain size. Genetics 165, 2063–2070.

Zilhão, J., Angelucci, D., Badal-García, E., d'Errico, F., Daniel, F., Dayet, L., Douka, K., Higham, T.F.G., Martínez-Sánchez, M.J., Montes-Bernárdez, R., Murcia-Mascarós, S., Pérez-Sirvent, C., Roldán-García, C., Vanhaeren, M., Villaverde, V., Wood, R., Zapata, J., 2010. Symbolic use of marine shells and mineral pigments by Iberian Neandertals. Proceedings of the National Academy of Sciences, USA 107, 1023–1028.

Chapter 2

Creativity and Complex Society Before the Upper Palaeolithic Transition

Clive Gamble

Faculty of Humanities (Archaeology), Building 65A, Avenue Campus, University of Southampton, Southampton SO17 1BF, Clive.Gamble@soton.ac.uk

My aim in this paper is to examine the proposition that creativity in the Palaeolithic is not to be measured by things alone. Instead I will draw attention to the importance of emotions in the social lives of hominins and especially their role as a mechanism for innovation. Furthermore, I will argue that the recent interest in Pleistocene demography to explain the presence, rarity or absence of artefacts that symbolise, for archaeologists, innovation and hence creativity needs to be re-thought. The issues considered here are

1. How should the hominin mind be modelled?
2. How many minds were needed for hominin creativity to evolve?
3. What is the role of the social?
4. How does creativity relate to social complexity?

But first we need to define the problem.

2.1. A DEFINITION

I will argue that viewing creativity as nothing more than a synonym for adaptation through innovation obscures the role of the emotions and presents us only with half-formed hominins. While new artefacts such as boats and bows undoubtedly contributed to the survival and reproductive success of both individuals and groups, this adaptive position ignores the basis of cultural interaction which depends on our sensory responses to both materials *and* people. Creativity is an embodied act and it is a social act. To regard creativity as a mental or cognitive process alone biases any discussion towards considering only novelty and the adaptive conditions of its acceptance. Moreover, archaeological data then serve little purpose other than as time markers on the long journey to creative complexity, the way stations of hominin, anatomically modern human and fully modern human marked by more new things. But the lack of novelty does not necessarily imply a dearth of creativity. Hominins could have been creative in terms of how they engaged with each other even while they repetitively made the 100-millionth handaxe of the archaeological record. Also, those daily forms of social engagement, while no doubt enlivened by the very first piece of carved reindeer antler, did not become any less creative because they were now overshadowed by a new thing. Rather, the emotional impact of those interactions was *enhanced* by innovation. Creativity is therefore a mechanism of social engagement that amplifies the sensory mechanisms by which bonds are formed. Moreover, these relationships can be extended in time and space because they involve both people and things.

2.2. MODELLING THE HOMININ MIND

At a time when the death of the mind has been pronounced (Gosden, 2010) it may seem contrarian to highlight its importance for creativity. However, dead or alive it is the nature of the intellectual baggage which comes with concepts of the mind that is central to understanding the issue.

Gosden's target is twofold. The easy part is a tilt at the Cartesians with their internal model of a mind that works in isolation from the world; a rational, problem-solving mind that sees what needs to be done to maximise reproductive benefits, minimise costs and so turn an evolutionary profit by devising new adaptations. Mithen's (1996) influential model of the hominin mind follows this pattern with its five modules of intelligence (social, natural history, language, technical and general). Cognitive fluidity provided the all-important linkages between the modules and it is this capacity, and the novel associations it produced, that drove creative evolution. For example, horses became not only good to eat (natural history intelligence module) but now, due to cognitive fluidity, good to symbolise the clan to which the eater belonged (social intelligence module).

This rational, internal model of the mind is used by most archaeologists (see Gamble, 2010 for discussion). For example, it is manifested in the notion of external storage where the symbolic content of the mind is offloaded and stored outside the individual (Renfrew and Scarre, 1998). As a result, this part of the mind is separated from both the body of the individual and the world in which they live. A mind of symbolic things is created and then re-accessed by rationally considering its usefulness from the vantage point of an independent mind of neuron-linked memories. But to achieve this utility the common criticism of the Cartesian mind, as one populated by ever-smaller homunculi all observing each other, is often ignored by supporters of external storage. The homunculus now takes up residence in the mind of the symbolic object; it gazes out through the eyes of the figurine or from the 8-GB memory stick in your pocket.

Gosden's second target is the model of the mind as a distributed or extended cognition (Clark and Chalmers, 1998; Dunbar et al., 2010a). While sympathetic to the notion that cognition does not stop at the skin but forms a mind based on relational rather than rational observations, Gosden feels that many of its supporters are "still reacting to older thoughts about mind, saying that the mind exists not in the head but in the interaction between people and the world (Gosden, 2010: 39)". Rather than always acting on the world, he urges those interested in distributed cognition to see the world acting back on the brain and the body as an equal partner. So, for a hominin wielding a large cutting tool it would be as correct to say that their brains are a part of the extended handaxe as vice versa. Taken together, the brain, the body and the biface (the last a proxy for the surrounding environment) might be described collectively as a mind, but due to the excess historical baggage associated with the concept of mind we inevitably give priority to the grey matter in our skulls and so downgrade the equal importance of the body and the world.

My starting point in the discussion of hominin creativity and innovation is therefore that our cognition is relational rather than rational (Coward and Gamble, 2008; Grove and Coward, 2008). I will follow Gosden's lead and look for hominin creativity not just in the appearance of symbolic items that were made and manipulated but rather as a wider consideration of what being en-minded entails. Therefore, the Palaeolithic mind is no different from the modern mind, or the mind of many social animals (Barrett et al., 2007), in that it consists of body (senses and emotions) and things (both animate and inanimate) in reciprocal social partnerships rather than being directed solely by mental instructions sent from the seat of reason.

2.3. THE NUMBER OF MINDS AND THE EVOLUTION OF CREATIVITY

Adopting a distributed model of mind leads to an early casualty in the form of a symbolic revolution as an instance of species-level creativity, whether it was led by *H. heidelbergensis*, ~400 ka ago (Mania and Mania, 2005), *H. neanderthalensis*, ~60 ka ago (Zilhão et al., 2010) or *H. sapiens* ~40 ka ago (Klein, 2008). Irrespective of what was made, the relational structure of any distributed mind is the same. It applies to the small hominin brains and the stone tools of Gona at 2.5 Ma (Semaw et al., 1997) as it does to any chimpanzee with its termite probe (McGrew, 1992) or the brain of a New Caledonian crow and the varied implements it manufactures to obtain food (Hunt, 1996).

Hominin, primate and avian species who clearly innovate would seem to provide data for a broader evolutionary account of creative culture. Leading the way has been the dual-inheritance theory that examines the reproductive benefits for innovative adaptations and the conditions under which they become fixed in the cultural repertoire. For example, Shennan's (2001) mathematical model of cultural evolution claims that the consequences of innovation appear to be far more successful in larger rather than smaller populations. Larger populations have an advantage because, as Shennan (2001: 12) argues, the deleterious sampling effects of cultural innovations decline as population size increases. Belonging to a larger population brings its members benefits that enhance evolutionary fitness. The problem with small populations is that innovations which do not contribute significantly to reproductive success, and that are less attractive to imitate, can still be retained rather than selected out of the cultural repertoire (Shennan, 2001: 12). According to this model, creativity and innovation are therefore spoken of as adaptations and maladaptations. They can be analysed via cost–benefit analysis within the evolutionary context of enhanced reproductive fitness.

In the dual-inheritance theory there are two principal routes to such innovations. In vertical transmission, traits are passed through the varied mechanisms of social learning in a system of generational descent within a population. This produces phylogenies of cultural traits. In horizontal (or geographic) transmission, phylogenesis is replaced by ethnogenesis. Here it is the varied mechanisms of cultural diffusion that account for the process of change. In both instances, social selection has a role to play, as do genetic drift and the founder effect. All of these mechanisms, it is argued, are density dependent. In other words, it is the size of the population in which transmission takes place that determines the selective advantage of a) innovating and b) retaining that novel adaptation in either the current population (ethnogenesis) or next generation (phylogenesis).

The particular aspect on which to focus in the application of evolutionary theory to culture change is the role of demography. This helps explain why dual-inheritance models often gravitate for proof of concept to the small-world societies of hunters and gatherers. The Tasmanians, taken as a paradigmatic cultural isolate, have received especial attention ranging from Jones' (1977) assessment that they suffered a "slow strangulation of the mind" to Henrich's (2004) analysis of their cultural behaviours as an example of what he terms cumulative adaptive evolution.

These models have been applied to wider archaeological problems such as the paucity of symbolic artefacts during the initial colonisation of Australia in the Pleistocene. For example, following Cullen's (1996) metaphor of too-few-trees-[or minds] in-the-forest, Brumm and Moore (2005) argue that the rarity of such artefacts, compared to Europe and some parts of Africa, was due to low population densities during initial colonisation. The implication is that social learning was constrained by the low frequency of contact and the small numbers involved. The implication is that both Pleistocene Australians and Europe's much older *Homo heidelbergensis* would have had the capacity for symbolic creativity, as demonstrated by the rare artefacts interpreted as such (Mania and Mania, 2005), but they lacked the demographic densities to achieve successful transmission, by either vertical or horizontal modes. The same argument might also be applied to the discovery of some ochre-stained shells in Iberia dated to 50 ka that are associated with Middle Palaeolithic artefacts at Cueva de los Aviones and Cueva Antón (Zilhão et al., 2010). In brief, the capacity for symbolic thought and its material expression was present among Neanderthals but they were simply too few and too scattered to "fix" these traits in their cultural repertoire, as those traits arose. Brumm and Moore's (2005: 189) claim is a persuasive instance of the too-few-trees/minds argument and where "the absence of evidence for repeated patterning in symbolic behaviour cannot by itself be taken as evidence for the absence of behavioural modernity among past people" since it applies to humans and hominins.

In a recent paper, Powell et al. (2009) have applied the demography of cultural transmission to understanding the different times at which regional populations made the move to a symbolically organised world, what they refer to as the heterogeneous spatial and temporal structuring of the Upper Palaeolithic (UP) transition. Their concern is with the appearance, in quantity, of a range of items that might be regarded as innovations in this sphere of hominin life. These include the prepared core technique of making stone blades, varied forms of ornamentation and ochre use, functional and ritual use of bone, antler and ivory artefacts, grinding and pounding tools, musical instruments, long-distance trade and composite tools generally.

Their analysis proceeds by benchmarking the heterogeneous transition with the evidence from Europe where the Upper Palaeolithic appeared ~45 ka ago. Using estimates of effective population size (the number of breeding females) obtained from coalescent theory applied to mtDNA data (Atkinson et al., 2007) they derive a density figure for the inhabitable European landmass that is then applied to other regions (Table 2.1). However, their claim (Powell, Shennan, and Thomas 2009:1301) that the age estimates (Table 1) are supported by archaeological data is not accurate.

In particular, the discovery of *Nassarius* beads at Grotte des Pigeons, Taforalt, in Morrocco (Bouzouggar et al., 2007), dated by OSL to 82 ka, parallels the finds made in South Africa at the Blombos Cave (D'Errico et al., 2005). Powell, Shennan and Thomas regard the Blombos evidence as indicative of the UP transition, but seem prepared to wait for a later date in North Africa and the Middle East, even though a key artefact, *Nassarius* beads, is the same age at either end of the continent.

TABLE 2.1 The Upper Palaeolithic (UP) Transition Based on a Median Estimate for Effective Population Size of 2905 at 45 ka When Europe Made the Transition to the UP. Powell et al. (2009) Estimate the Transition Time in Other Regions of the Old World When this Threshold was Reached (iii)

	(i)	(ii)	(iii)
	Million km^2	Density × 10^{-4} km^2	Estimated Date, ka BP, of the UP Transition Data from Powell et al. (2009)
Europe	~8.883	~3.2714	45
Sub-Saharan Africa	~24.270		~101
Middle East and North Africa	~13.588		~40
Southern Asia			~52
Northern and Central Asia			~40

Further complications for testing their model arise with data from a Middle Palaeolithic context at the Skhul Cave in Israel, where pierced shell has now been dated to 130 ka (Vanhaeren et al., 2006). However, all of these estimates fall far short of the ages of 160 ka obtained from Pinnacle Point, South Africa, for the manufacture of micro-lithic tools, worked ochre and utilised shellfish (Marean et al., 2007). These traits figure widely in discussions of the transition to modern humans and at Pinnacle Point occur during MIS6, a severe cold stage, when population densities according to Shennan (2001: Figure 4) would have been very low because of the prevailing climate.

A further problem with the model of Powell et al. (2009) concerns the southern Asian data. They believe their estimate of ~52 ka is considerably older than "the first archaeological evidence for modern behaviour at ~30 ka" (2009: 1301). Yet based on the application of coalescent theory to mtDNA, this region had the fastest growth of effective population outside of Africa – growth that is also estimated to begin at ~52 ka (Atkinson et al., 2007: Figure 1). Furthermore, the mtDNA study shows that between 45 ka and 20 ka more than half of the global human population lived on the Indian subcontinent and the Thai and Malay peninsulas and their exposed continental shelves (Atkinson et al., 2007: 470). If we applied these demographic estimates to the Powell et al. (2009) model of cultural innovation we should expect the generation of abundant novel symbolic and cultural items in this region over this time period. But even allowing for poor preservation of organic remains, the region compares unfavourably in this regard with southern Africa, North Africa the Middle East and Europe.

Preservation of organic materials is not, however, the issue elsewhere in southern Asia. The Niah Cave, Borneo, contains an anatomically modern human skull, dated to either 37–35 ka or 44–40 ka BP, and abundant, often cut marked, animal bone. The archaeology is summed up by Barker: "Early modern humans in Sundaland may not have exhibited some of the classic indicators of modernity as defined in the European Aurignacian, such as refined blade and bladelet technology, body ornamentation, and mobiliary and parietal art, but their subsistence practices and engagement with the landscape were of demonstrable socio-economic complexity" (Barker et al., 2007: 259). This complexity included mammal and fish trapping, projectile technology, tuber digging, plant detoxification and forest burning. Roughly contemporaneous evidence from starch grains and carbonised plant remains in highland Papua New Guinea suggests the novel exploitation at high altitude of *Pandanus* nuts and yams in the interval 49–36 ka (Summerhayes et al., 2010). These finds are among the earliest for the colonisation of the palaeocontinent of Sahul.

The archaeological evidence from both regions suggests other measures of the UP transition and which in the form of landscape management (Borneo) and population dispersal into unoccupied habitats (PNGs) both speak of responses to, and consequences of, demographic growth. These innovative and creative acts are at least as worthy of consideration as carved bone figurines, painted cave walls and blade technologies. Yet, as shown so clearly by Powell et al. (2009), it is these symbolic artefacts, redolent of a rational Cartesian mind, that steal the show. By contrast, the evidence from landscape use of engagement with the environment in novel ways receives lesser billing.

2.4. SOCIAL BRAINS AND ACTIVE PERSONAL NETWORKS

A further problem with the demographic approach to innovation and creativity is that it says little about the composition and cognitive demands of group living. For Shennan (2001) it is the number of minds that matters, rather than the various ways they were linked together to form social relationships and the cognitive changes that were necessary to stabilise larger numbers. An indication of these key issues is provided by the social brain model that has revealed a strong correlation between brain and community size (Aiello and Dunbar, 1993; Gamble et al., 2011). This is what Roberts (2010) refers to as an individual's active network. Predicted community sizes vary from 80 for the australopithecines to 100 for early *Homo*. After 600 ka, when encephalisation increased significantly (Rightmire, 2004), the active networks range from 120 to 150 for a number of large-brained hominin species.

Time forms a major constraint for these larger active networks. Primates use finger-tip grooming to establish social bonds on a daily basis among all the levels and groups in their active networks. But this result cannot be achieved in the time that is available to large-brained hominins because of their increased number of social partners. Instead, language represents one solution to this time constraint (Aiello and Dunbar, 1993) as does the elaboration of material culture (Gamble, 2009). But not only are these community sizes larger, they are also less stable. They fragment under fission and fusion and the need to respond to the seasonal availability of food and water. The point about larger community sizes is not the additional 50 members between early and late species of *Homo* but rather the greater number of subunits that have to be incorporated. To give some idea of the exponential nature of the problem, Kephart (1950) calculated that the potential number of social relationships within a group of 5 people was 90 and that this rose to 7,133,616 with 15 group members (Gamble, 1999: 429 footnote 7), suggesting that ways have to be found to manage the cognitive load arising from social interaction.

Hierarchies are one way to achieve this (Johnson, 1982). For example, Roberts (2010) has examined the inclusive structure of human personal networks that form networks of 5 (support clique), 15 (sympathy group), 50 (band) and 150 (the active network). This pattern has been further studied by Zhou et al. (2004) as a geometric series with a scaling ratio of three. Therefore, as the size of a hominin's active network increased from early to late *Homo,* so too did the number of networks, or local groups that had to be integrated into a stable social whole. This required solutions to constraints on cognitive load, allowing hominins to interact at sufficient intensity but for shorter periods of time, to maintain community cohesion. Therefore, brain size of fossil hominins is a proxy for increasing social complexity (Gamble et al., 2011).

2.5. CREATIVITY, THE SENSES AND SOCIAL COMPLEXITY

Social complexity is the issue that emerges from a consideration of the linkage between brain and group size during human evolution. Consequently, it is social complexity that should frame the discussion of creativity and innovation in the Palaeolithic. Furthermore, the notion of social complexity is not restricted solely to the Upper Palaeolithic with its artistic outbursts in some, but by no means all, regions of the world occupied by *H. sapiens*. Complex social behaviour, as I have defined it in this paper, has to be acknowledged as a property of all large-brained hominins after 600 ka, including *Homo heidelbergensis*, Neanderthals, late *H. erectus* as well as *H. sapiens*. In the last case, this is irrespective of their classification as anatomically, rather than behaviourally, modern — a judgement based entirely on the artefacts with which they are associated. What then emerges as interesting from the archaeology of the last 600 ka is that *things* do not get appreciably more complex until long after social complexity has occurred, and, even then, as Brumm and Moore (2005) point out, it does not have to happen.

As an alternative, I suggest that human creativity has never been limited to the invention or modification of physical objects. Hominins have used creativity to solve the constraints of larger community size through amplifying the intensity of social interaction – in other words, to create more lasting bonds that bind and define.

But can archaeologists measure creativity other than by things? This is where emotions offer an insight. Instead of arguing that pride, happiness, disgust or shame accompanied the making of flint tools or carved bones and ochre, I will consider the role of emotions as a resource that can be amplified to strengthen social bonds. Emotions leave no archaeological evidence, or at least that is what our Cartesian training has taught us. However, since Darwin (1872) it has been widely acknowledged that emotions play a fundamental role in sexual selection and the creation of social bonds and relationships. For example, the investigation of primate society was made possible by the study of emotions that create bonds that bind (de Waal, 2006). By ignoring emotions and relying solely on things that survive to reconstruct social behaviour, Palaeolithic archaeologists present a picture of the social lives of large-brained hominins such as Neanderthals that is impoverished when compared to a small-brained primates like the Aye–Aye. This situation has always been anomalous and suggests archaeologists have been too constrained by their models not only of society (Gamble, 1999) but also of minds (Dunbar et al., 2010b).

Identifying individual emotions is not the issue, and our inability to see, let alone feel, anger in the materials of the past should not stop the enquiry. Instead, emotions can be classified into three broad sets: background (mood), primary (survival) and secondary (social) as shown in Table 2.2 (Damasio, 2000; Gamble, 2009; Turner, 2000). Of particular interest from an evolutionary perspective are

TABLE 2.2 Some Examples of a Hierarchy of Emotional States (Damasio, 2000; Turner, 2000). These can be Further Elaborated by Recognising Levels of Intensity e.g., Downcast — Gloomy — Sorrow (Gamble, 2009: Table 2.2). The Use of Emotions to Amplify Social Life Depends on Cognitive Advances Such as Theory of Mind. None of these Emotions has a Material Proxy for the Archaeologist to Recover but that does Not Diminish their Importance

Theory of Mind				
Social or secondary emotions	Guilt	Envy	Shame	Pride
Common primate ancestry				
Primary or survival emotions	Fear	Anger	Happiness	Sorrow
Affect and mood				
Background or place emotions	Calm	Safety	Fraught	Haunted

the social emotions – shame, envy, jealousy and pride to name a few. They are secondary in that they require a theory of mind that acknowledges another point of view. Without that perspective there can be no guilt or moral basis to social interaction. The possible outcomes from social encounters become based less on the experience of aggression or pacification and more on codes established by patterns of social trust that recognise multiple viewpoints. Such a theory of mind is not possessed by any primate and by extension any of the small-brained hominins. Social emotions that required such a cognitive ability would, however, have been commonplace after 600 ka, and not restricted to modern humans alone (Gamble et al., 2011).

The essential point about emotions is that, together with materials, they are the scaffolding that surrounds the construction of social life (Gamble et al., 2011: Figure 2). Materials and emotions are the resources that combine with brains to form the distributed mind I described earlier. All social life is scaffolded around them. Stones, wood, fibre, water, fire, clay and all the multifarious materials of the world are transformed by practices into familiar patterns and traditions, social practices and forms. In the same way, the senses of the body are channelled into definable emotional states to which our bodily chemistry responds. The opportunity for the amplification of emotions to underpin more complex social interaction is therefore possible. Mechanisms such as social laughter, singing, dancing, sports and making music all increase the emotional content of performances between people and so contribute to enhanced social bonds (see papers in Dunbar et al., 2010a).

2.6. CONCLUSION

What is significant for the discussion of creativity and innovation is that this core of the distributed mind – neurons, materials and emotion – is available for amplification. Such intensification of the core serves a single purpose which is to define social bonds in order to promulgate social trust beyond immediate knowledge and face-to-face experience.

The argument of "too-few-trees/minds-in-the-forest" has, as we have seen, been applied to both the lack and paucity of behaviourally significant traits (symbolic things) as well as to their abundance during transitions. Powell et al. (2009: 1299) are correct in saying that not only does the transition need explaining but also its heterogeneous spatial and temporal structuring. But their model is not supported by archaeological evidence *as they choose to use it*. For Brumm, Moore and Zilhao, small demographic size is also an explanation for what *they know must be there* despite the scant evidence. In the case of Australia this is because no one disputes that it was colonised by modern humans in the Pleistocene. To argue otherwise is to return to colonial judgements about the intelligence of people at the uttermost ends of the Earth (Jones, 1992). While for the Iberian Neanderthals, the assumption that they were not capable of being modern humans becomes a comparable racial slur. The alternative that I have discussed in this paper requires two things: change the model of the Palaeolithic mind and embrace the power of emotions as a creative agent for change in human evolution.

ACKNOWLEDGEMENTS

The research in this paper was supported by the British Academy Centenary Project *From Lucy to language*: *the archaeology of the social brain*.

REFERENCES

Aiello, L., Dunbar, R., 1993. Neocortex size, group size and the evolution of language. Curr. Anthropol. 34, 184–193.

Atkinson, Q.D., Gray, R.D., Drummond, A.J., 2007. mtDNA variation predicts population size in humans and reveals a major southern Asian chapter in human prehistory. Mol. Biol. Evol. 25, 468–474.

Barker, G., Barton, H., Bird, M., Daly, P., Datan, I., Dykes, A., Farr, L., Gilbertson, D., Harrisson, B., Hunt, C., Higham, T.F.G., Kealhofer, L., Lewis, K.J.H., McLaren, S., Paz, V., Pike, A., Piper, P., Pyatt, B., Rabett, R., Reynolds, T., Rose, J., Rushworth, G., Stephens, M., Stringer, C., Thompson, J., Turney, C., 2007. The 'human revolution' in lowland tropical Southeast Asia: the antiquity and behaviour of anatomically modern humans at Niah Cave (Sarawak, Borneo). J. Hum. Evol. 52, 243–261.

Barrett, L., Henzi, P., Rendall, D., 2007. Social brains, simple minds: does social complexity really require cognitive complexity? In: Emery, N., Clayton, N., Frith, C. (Eds.), Social Intelligence: From Brain to Culture. Oxford University Press, Oxford, pp. 123–146.

Bouzouggar, A., Barton, N., Vanhaeren, M., d'Errico, F., Collcut, S.N., Higham, T.F.G., Hodge, E., Parfitt, S., Rhodes, E., Schwenninger, J.-L., Stringer, C., Turner, E., Ward, S., Moutmir, A., Stamboul, A., 2007. 82,000 year old shell beads from North Africa and implications for the origins of modern human behavior. Proc. Natl. Acad. Sci. U S A 104, 9964–9969.

Brumm, A., Moore, M.W., 2005. Symbolic revolutions and the Australian archaeological record. Camb. Archaeol. J. 15, 157–175.

Clark, A., Chalmers, D.A., 1998. The extended mind. Analysis 58, 7–19.

Coward, F., Gamble, C., 2008. Big brains, small worlds: material culture and the evolution of mind. Philos. Trans. R Soc. Lond. B Biol. Sci. 363, 1969–1979.

Cullen, B., 1996. Social interaction and viral phenomena. In: Steele, J., Shennan, S. (Eds.), The Archaeology of Human Ancestry. Routledge, London, pp. 420–433.

Damasio, A., 2000. The Feeling of what Happens: Body, Emotion and the Making of Consciousness. Vintage, London.

Darwin, C., 1872. The Expression of the Emotions in Man and Animals, third ed. John Murray, London.

de Waal, F., 2006. Primates and Philosophers: How Morality Evolved. Princeton University Press, Princeton.

D'Errico, F., Henshilwood, C.S., Vanhaeren, M., van Niekerk, K., 2005. *Nassarius kraussianus* shell beads from Blombos Cave: evidence for symbolic behaviour in the Middle Stone Age. J. Hum. Evol. 48, 3–24.

Dunbar, R., Gamble, C., Gowlett, J.A.J. (Eds.), 2010a. Social Brain and Distributed Mind. Oxford University Press, Oxford Proceedings of the British Academy 158.

Dunbar, R., Gamble, C.S. Gowlett, J.A.J. (Eds.), 2010b. The Social Brain and the Distributed Mind. Oxford University Press, Oxford, pp. 3–15. Proceedings of the British Academy 158.

Gamble, C.S., 1999. The Palaeolithic Societies of Europe. Cambridge University Press, Cambridge.

Gamble, C.S., 2009. Human display and dispersal: a case study from biotidal Britain in the middle and upper pleistocene. Evol. Anthropol. 18, 144–156.

Gamble, C.S., 2010. Technologies of separation and the evolution of social extension. In: Dunbar, R., Gamble, C., Gowlett, J.A.J. (Eds.), Social Brain and Distributed Mind. Oxford University Press, Oxford, pp. 17–42. Proceedings of the British Academy 158.

Gamble, C.S., Gowlett, J.A.J., Dunbar, R., 2011. The social brain and the shape of the Palaeolithic. Camb. Archaeol. J.

Gosden, C., 2010. The death of the mind. In: Malafouris, L., Renfrew, C. (Eds.), The Cognitive Life of Things: Recasting the Boundaries of the Mind. McDonald Institute for Archaeological Research, Cambridge, pp. 39–46.

Grove, M., Coward, F., 2008. From individual neurons to social brains. Camb. Archaeol. J. 18, 387–400.

Henrich, J., 2004. Demography and cultural evolution: how adaptive cultural processes can produce maladaptive losses – the Tasmanian case. Am. Antiq. 69, 197–214.

Hunt, G.R., 1996. Manufacture and use of hook-tools by New Caledonaian crows. Nature 379, 249–251.

Johnson, G., 1982. Organizational structure and scalar stress. In: Renfrew, C., Rowlands, M., Segraves, B. (Eds.), Theory and Explanation in Archaeology: The Southampton Conference. Academic Press, New York, pp. 389–422.

Jones, R., 1977. The Tasmanian paradox. In: Wright, R.V.S. (Ed.), Stone Tools as Cultural Markers. Australian Institute of Aboriginal Studies, Canberra, pp. 189–204.

Jones, R., 1992. Philosophical time travellers. Antiquity 66, 744–757.

Kephart, W.M., 1950. A quantitative analysis of intragroup relationships. Am. J. Sociol. 60, 544–549.

Klein, R.G., 2008. Out of Africa and the evolution of human behaviour. Evol. Anthropol. 17, 267–281.

Mania, D., Mania, U., 2005. The natural and socio-cultural environment of homo erectus at Bilzingsleben, Germany. In: Gamble, C., Porr, M. (Eds.), The Individual Hominid in Context: Archaeological Investigations of Lower and Middle Palaeolithic Landscapes, Locales and Artefacts. Routledge, London, pp. 98–114.

Marean, C.W., Bar-Matthews, M., Bernatchez, J., Fisher, E., Goldberg, P., Herries, A.I.R., Jacobs, Z., Jerardino, A., Karkanas, P., Minichillo, T., Nilssen, P.J., Thompson, E., Watts, I., Williams, H.M., 2007. Early human use of marine resources and pigment in South Africa during the middle Pleistocene. Nature 449, 905–908.

McGrew, W.C., 1992. Chimpanzee Material Culture: Implications for Human Evolution. Cambridge University Press, Cambridge.

Mithen, S., 1996. The Prehistory of the Mind. Thames and Hudson, London.

Powell, A., Shennan, S., Thomas, M.G., 2009. Late Pleistocene demography and the appearance of modern human behavior. Science 324, 1298–1301.

Renfrew, C., Scarre, C. (Eds.), 1998. Cognitive Storage and Material Culture: The Archaeology of Symbolic Storage. McDonald Institute, Cambridge.

Rightmire, P., 2004. Brain size and encephalization in Early to Mid-Pleistocene *Homo*. Am. J. Phys. Anthropol. 124, 109–123.

Roberts, S.G.B., 2010. Constraints on social networks. In: Dunbar, R., Gamble, C., Gowlett, J.A.J. (Eds.), Social Brain and Distributed Mind. Oxford University Press, Oxford, pp. 115–134. Proceedings of the British Academy 158.

Semaw, S., Renne, P., Harris, J.W.K., Feibel, C.S., Bernor, R.L., Fesseha, N., Mowbray, K., 1997. 2.5 million-year-old stone tools from Gona, Ethiopia. Nature 385, 333–336.

Shennan, S., 2001. Demography and cultural innovation: a model and its implications for the emergence of modern human culture. Camb. Archaeol. J. 11, 5–16.

Summerhayes, G.R., Leavesley, M., Fairbairn, A., Mandui, H., Field, J., Ford, A., Fullagar, R., 2010. Human adaptation and plant use in Highland New Guinea 49,000 to 44,000 years ago. Science 330, 78–81.

Turner, J.H., 2000. On the Origins of Human Emotions: A Sociological Inquiry Into the Evolution of Human Affect. Stanford University Press, Stanford.

Vanhaeren, M., d'Errico, F., Stringer, C., James, S.L., Todd, J.A., Mienis, H.K., 2006. Middle Palaeolithic shell beads in Israel and Algeria. Science 312, 1785–1788.

Zhou, W.-X., Sornette, D., Hill, R.A., Dunbar, R., 2004. Discrete hierarchical organization of social group sizes. Philos. Trans. R Soc. Lond. B Biol. Sci. 272, 439–444.

Zilhão, J., Angelucci, D.E., Badal-Garcia, E., d'Errico, F., Daniel, F., Dayet, L., Douka, K., Higham, T.F.G., Martinez-Sánchez, M.J., Montes-Bernárdez, R., Murcia-Mascarós, S., Pérez-Sirvent, C., Roldán-Garcia, C., Vanhaeren, M., Villaverde, V., Wood, R., Zapata., J., 2010. Symbolic use of marine shells and mineral pigments by Iberian Neandertals. Proc. Natl. Acad. Sci. U S A Early Edition.

Chapter 3

North African Origins of Symbolically Mediated Behaviour and the Aterian

Nick Barton[1] and Francesco d'Errico[2,3]

[1] Institute of Archaeology, University of Oxford, 36 Beaumont Street, Oxford OX1 2PG, UK
[2] CNRS UMR 5199 PACEA, Université Bordeaux 1, avenue des Facultés, F-33405 Talence, France
[3] Department of Archeology, History, Cultural Studies and Religion, University of Bergen, Norway

3.1. INTRODUCTION

One of the liveliest topics of debate in evolutionary studies today concerns the emergence of modern behaviour in humans. In spite of general agreement that physically modern *Homo sapiens* arose in East Africa at c. 200,000 years ago (200 ka), enormous controversy still surrounds whether modern cognition and associated innovations happened as a direct consequence of the emergence of our species, and whether knowledge spread from one centre or developed autonomously at different times and in different parts of Africa. Equally controversial is whether quintessentially modern behaviours such as the ability to act and think symbolically were unique to this new species or evolved independently amongst widely separated and, to a large extent, genetically distinct populations.

In this chapter, we shall side step some of the latest concerns over semantic issues dealing with "behavioural modernity" and "behavioural variability" in early hominins (Shea, 2011; Marean, 2010) and concentrate instead on the nature and timing of a few key innovations in the cultural record of North Africa during the Middle Palaeolithic/Middle Stone Age (henceforward MP/MSA). This is a region that has so far received scant attention in debates regarding the understanding of behavioural change in *H. sapiens*. In particular we focus on the technocomplex known as the Aterian and discuss whether the first appearance of explicitly symbolic objects in this context reflected a stage of rapid acceleration in cultural change across this region. A related issue concerns the question of what if any external factors may have given rise to innovations in technology and symbolic material culture and, finally, why these developments may have lacked longevity and ended in cultural loss as they probably also did in other parts of Africa.

3.2. SYMBOLISM AND HUMANS

Defining what we mean by "symbolism" and being able to successfully demonstrate it from material evidence in the early human record present obvious practical and theoretical challenges, not least the difficult question of how this behaviour might have manifested itself archaeologically. For some authors, the attribution of specific meanings to conventional signs – a fundamental step in the development of symbolic thinking – may be present in nascent form in various primate and nonprimate societies (Byrne, 1995; Rumbaugh and Washburn, 2003; Tomasello et al., 2005) and thus may not be unique to our species. On the other hand, despite the fact that chimpanzees clearly have a capacity for culture and the ability to transmit traditions (Whiten, 2005), these cannot be compared to the sophisticated creation of symbols, and incorporating them in material culture or displaying them on the body as seen in modern peoples.

Thus, symbolic activity or the ability to share, store and transmit coded information within and across groups (d'Errico and Stringer, 2011) would appear to be intrinsic to all modern humans. Moreover, it can be shown that symbolically mediated behaviours play a fundamental role in the creation of modern cultures (Deacon, 1997; Donald, 1991; Rossano, 2010), crucially in creating and maintaining technical and social conventions, beliefs and identities. Although such behavioural features may leave little or no direct archaeological traces behind them, we would suggest that the most durable signs of this behaviour and the most likely proxies would come from personal ornamentation in the form of body paints or small portable items displayed on the body; art, including abstract and figurative representations; and indirect evidence for ceremonies or rituals arising from complex treatment of the dead.

3.3. NORTH AFRICA DURING THE MSA AND THE ATERIAN

Until recently North Africa would rarely have featured in debates on early symbolism and the emergence of modern human origins. This can be explained partly by the paucity of published data on key fossil and archaeological sites and the generally impoverished nature of the dating evidence. In addition, the MP/MSA of this region has been caricatured as consisting of a long, and largely unchanging, sequence of lithic industries with key innovations arising later there rather than elsewhere and therefore of marginal relevance to broader questions concerning the tempo and scale of changes that feature in discussions about behavioural modernity. As such, North Africa has frequently been treated as a "cul de sac", leaving it overshadowed by the more prolific records of East and sub-Saharan Africa.

Of key significance in the revival of interest in the question of early symbolism has been the reassessment of the Aterian. This MP/MSA industry is recorded across much of the Sahara from the Atlantic coast of Morocco to Egypt and Sudan in the east (Caton-Thompson, 1946) but apparently excluding parts of north east Africa (Clark, 1993) and the Nile Valley, where other MP/MSA complexes had been recognised (Wendorf and Schild, 1976; Kleindienst, 1998; Van Peer, 1991). Based on typologically diagnostic pedunculate (tanged) points, this technology was first recognised at the Eckmuhl Cave in Algeria (Carrière, 1886), and given the formal name *Atérien* (Aterian) by Reygasse (1921–1922) after the site of Bir-El-Atir, south of Tébessa, in Algeria, though the industry from this eponymous location (now known as Oued Djebbana) was not described until much later (Caton-Thompson, 1946; Morel, 1974 a,b). According to Caton-Thompson, the Aterian had several technological variants but was nevertheless homogeneous enough to be considered a single, very widespread, cultural entity that had its origins in equatorial Africa (Caton-Thompson, 1946). Since then others have disputed its roots, suggesting the Aterian was broadly Maghrebian (e.g., Debénath et al., 1986) or had antecedents further east (Kleindienst, 1998) or was spread by anatomically modern human populations from an East African origin (McBrearty and Brooks, 2000). The latter has also provoked much disagreement and has been problematic because of the relatively few human remains found at Aterian sites. Hublin's suggestion (2000) that robust modern *H. sapiens* were present in north-west Africa even before the Aterian is also supported by recent studies of 50 hominin teeth from a number of different Aterian sites that reveal cusp patterns similar to those on the Qafzeh-Skhūl humans, as well as the modern human sample from Pestera cu Oase in Romania (Hublin et al., in press). Amongst other specimens purportedly of Aterian type is a partial skull from Dar es-Soltan II that may date to about 80,000 ka (Balter, 2011).

According to 19 of the recorded facial landmarks, it shows closest similarities with those of the Near East populations (Harvati and Hublin, in press), but, nevertheless, retaining some archaic features in morphology (Stringer, pers comm). Thus, it seems likely though not proven that the Aterian humans were related to anatomically moderns.

The antecedents for the Aterian in the Maghreb are not well understood. For a long time it was considered by European prehistorians to have emerged out of the Mousterian (Balout, 1965; Camps, 1974). It was sometimes referred to as a "Moustérien à outils pédonculés" (Reygasse, 1922) or "Mousterian with stemmed points" (Pond et al. nd cited in Kleindienst, 1998) or simply as a cultural facies of the Mousterian (Balout, 1965; Tixier, 1967). Even accepting there is such a thing as a Maghrebian Mousterian (see Kleindienst, 1998), there are very few sites where an unambiguous stratigraphic relationship with the Aterian is preserved (Bouzouggar and Barton, in press). Rare exceptions to this pattern include the caves of Rhafas and Ifri n'Ammar where MP/MSA material without pedunculates underlies typically Aterian assemblages (Wengler, 1997; Richter et al., 2010). Initial descriptions of the Mousterian layer 3b at Rhafas suggest that the Mousterian at this site was rich in flakes with mainly faceted platforms struck from preferential Levallois flake cores, and with a tool assemblage dominated by side-scrapers and very few end-scrapers (Wengler, 1997, Table 2). At Ifri n'Ammar, the Mousterian from the oldest Middle Palaeolithic layer (Lower OI: *Occupation Inférieure*) is separated from the overlying Aterian layer by a 0.2-m-thick, sterile deposit and contains side-scrapers, denticulates and notched pieces (Richter et al., 2010; Nami and Moser, 2010). At Djebel Irhoud, layer 18 has provided fossils of *H. sapiens* and contains a predominantly Levallois industry, described as Mousterian, with side-scrapers and only rare blades (Hublin et al., 1987). The assemblage may be atypical or biased due to the high proportion of certain tools but the total absence of pedunculates and end-scrapers implies that there was no Aterian at this site.

The definition of the Aterian most generally accepted is that of Tixier (1967). His classification is largely based on the industry from Oued Djebbana, Algeria, later elaborated by Morel (1974a,b). He described the Aterian as a Levallois industry with a laminar or blade-like debitage showing a high proportion of faceted butts. The tools include side-scrapers and points, with a predominance of end-scrapers, mostly on the ends of blades and sometimes making up 20% of the total tools. Another important category is pedunculate tools presenting a tang at their proximal end (Fig. 3.1). These often make up 25% of the tool assemblage, and he also drew particular attention to the bifacial thinning of the tanged ends (Tixier, 1967, 795). Together with the end-scrapers, the pedunculates are probably the most characteristic components of the Aterian and provide a clear

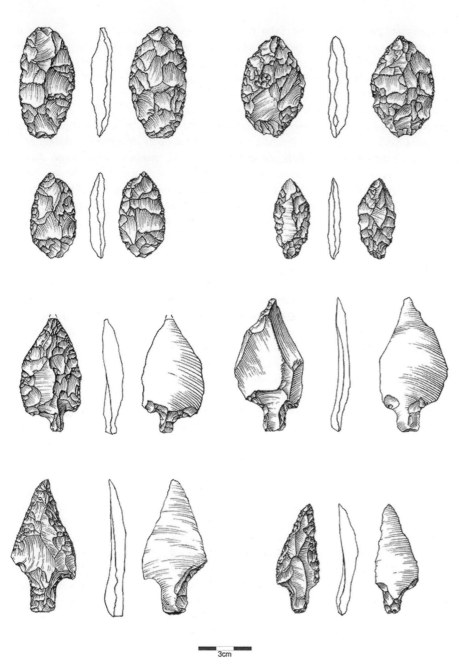

FIGURE 3.1 Aterian bifacial foliate points and pedunculates, Oued Djebanna, Algeria. *Drawings by Mathieu Leroyer*

distinction with other "Mousterian" assemblages. In addition to Tixier's scheme, Bouzouggar and Barton (in press) have signalled two other common attributes of Maghreb Aterian assemblages. The first concerns small centripetal Levallois and discoidal cores that are found in a large number of assemblages in Morocco and Algeria including Tit Mellil, Aïn Fritissa, Oued Djebbana, Contrebandiers, El Aliya, El Mnasra and El Harhoura 2 (Antoine, 1938; Tixier, 1958–1959; Bouzouggar, 1997; Bouzouggar et al., 2002; Nespoulet et al., 2008a). The small Levallois flakes were suitable for making pedunculates (Bouzouggar, 1997). The second category is represented by bifacial foliate points (Fig. 3.1). These distinctive tools may vary in size but are predominantly made on laminar flakes or blades and retouched on both faces. Invasive flaking on the ventral surface served to thin and straighten the tools. Although particular attention has been paid in the past to tangs as evidence for socket hafting (Alimen, 1957; Clark, 1982; Kleindienst, 1998), it is clear that the bifacial thinning of foliates must have served a similar purpose too.

Thus, one of the most striking features of the Aterian lithic technology is the co-occurrence of proximally thinned pedunculate tools and bifacially flaked foliate points. The appearance of these items, and especially the thin, all-over flaked points, is no doubt related to a sophisticated projectile technology. The overall symmetry and thinness of these tools, some no larger than a leaf-shaped arrowhead, may even suggest the existence of very early archery techniques rather than spear throwing. It is no coincidence that their presence has been recorded in association with predated ungulate faunas. At Taforalt, the first occurrence of pedunculates and foliate points is recorded in intra-MIS 5 layers that have also produced caprids and equids with indications of more steppic grassland conditions (Bouzouggar et al., 2007). A similar situation is reflected at Ifri n'Ammar in the *Occupation Inférieure*, which in addition to equids also includes White Rhinoceros (*Ceratotherium simum*), a specialised grazer (Hutterer in Nami and Moser, 2010).

3.4. A REVISED CHRONOLOGY FOR THE ATERIAN

Following the initiation of new excavation programmes in Morocco (e.g., INSAP-Oxford Caves Project begun in 2000 and led by A. Bouzouggar and N. Barton), and using multi- and single-grain OSL (optically stimulated luminescence), uranium-series, and TL (thermoluminescence) dating techniques, the Aterian can be shown to occupy a much longer time span than previously thought. Currently, the oldest plausible evidence for this technology can be found within the Upper *Occupation Inférieure* at Ifri n'Ammar which has been TL dated to 145 ka (Richter et al., 2010), though as has already been noted Mousterian artefacts have also been recovered from the Lower *Occupation Inférieure* which has associated dates of around 171 ka. Further west on the Atlantic coast in the Témara district of Rabat are cave sites associated with raised beach deposits of likely MIS 5e age (Barton et al., 2009). At Dar es-Soltan I the earliest Aterian industry in Rhulmann's layer I has been OSL dated to around 114 ka BP (Barton et al., 2009), while at the nearby site of El Mnasra, Aterian layers are in the range of 105–111 ka, with an additional OSL date of 118 ka for the top of layer 13, though this has yet to been confirmed as Aterian (Schwenninger et al., 2010). The site of Contrebandiers, which also lies nearby, contains Aterian artefacts in layers 8–14 with layer 11 providing ages of between 100 and 121 ka (Schwenninger et al., 2010). All of these dates are consistent with the presumed MIS 5e age of the underlying fossil beach line.

The new dating of the Aterian challenges the hypothesis of an introduction from the east across the Sahara. According to the OSL chronology, the Aterian can presently be shown to be older in Morocco than in Tunisia and the Libyan Sahara and along its northern fringes. In Tunisia, an Aterian-like industry at Oued El Akarit has recently been dated to more than 70 ka (Roset and Harbi-Riahi, 2007). In the Libyan Sahara, the oldest dates come from Uan Tabu rockshelter in the Tadrart Acacus, where the upper layer of a deposit with Aterian artefacts gave an OSL age of 61 ka (Garcea, 2001, 2010). Similarly in the Jebel Gharbi area of north-western Libya the earliest Aterian is tentatively dated by U/Th at 64 ka at Ain Zargha (Ras el Wadi) (Barich and Garcea, 2008; Garcea, 2009). Elsewhere, in Cyrenaica, there is little published evidence for the Aterian except from open-air contexts (Barker et al., 2010). At the Haua Fteah, artefacts with some Aterian affinities were reported by McBurney from layers 30–31 (1967, Fig. V.10) which in broad terms would belong to the period of about 60–80 ka. Despite the emerging picture of generally older dates in the west, it would be unwise to rule out the possibility of sites of comparable age further to the east. For example, recent work in the Kharga Oasis, Egypt, has reported the presence of a solitary Aterian tanged point in silts of Palaeolake Jaja which are capped by a tufa deposit with uranium-series ages of 126 ± 4 ka Uyrs (Smith et al., 2004) and 122.6 ± 1 ka Uyrs (Kleindienst et al., 2009), although ESR results on snail shells in silts beneath the tufa gave a younger minimum age of 96.24 ± 2.5 ka yrs (Kleindienst et al., 2009).

If the spread of anatomically modern humans from East Africa into North Africa took place in MIS 7 or MIS 6, the question arises that they may have done so with a different material culture than represented in the Aterian. The earliest fossils of *H. sapiens* from North Africa come from Jebel Irhoud in Morocco and are dated to between 190 and 160 ka (Smith et al., 2007). These are associated with an MP/MSA industry of Mousterian affinities and with no reported examples of pedunculates or bifacial foliates. Similar human types recovered from the Haua Fteah in north-east Libya (Cyrenaica) were recovered in layers containing Levallois–Mousterian material and, in any case, are likely to belong within the late MIS 5 part of the sequence. The fact that the Aterian tools included examples that must have been hafted suggests that this adaptation emerged independently in Africa and Europe, and appeared in North Africa probably during MIS6.

3.5. INNOVATIONS IN THE ATERIAN

In terms of innovations, the earliest Aterian is partly defined by the appearance of pedunculate tools and end-scrapers and possibly bifacial foliates. This is true for the Upper *Occupation Inférieure* at Ifri n'Ammar where pedunculates first appear in layer 54 but only in combination with foliates in layers 50–52 at the very top of this unit, at around 145 ka (Nami and Moser, 2010, Fig. 147). At Dar es-Soltan I, both types are present in Aterian layers dating from about 114 ka (Barton et al., 2009). Apart from the lithic technology, several other important novelties have been recorded in the early Aterian of Morocco. These include the use of colourants at Contrebandiers (Fig. 3.2)

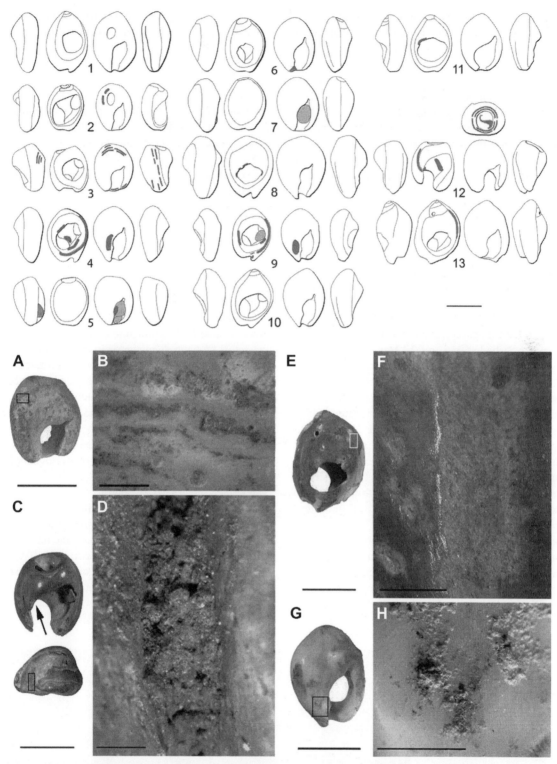

FIGURE 3.2 Nassarius shell beads from Grotte des Pigeons, Taforalt (photo by F. d'Errico) Top: areas in red identify the location of red pigment residues on *N. gibbosulus* shells from Grotte des Pigeons. Bottom: pigment residues on the ventral side (A-B, n.3), inside the bodywhorl (C-D n.12), on the dorsal side (E-F n.13) and close to the syphonal canal (G-H n.9). Black squares in A, C, E, G and arrow in C identify the area enlarged in the adjacent micrograph. Scale bar 1–13, A, C, E, G = 1 cm; B, D, F, H = 500 μm.

and at the El Mnasra Cave that are represented in Layer 7 of the latter site by small blocks of red pigment that exhibit subparallel striations probably produced by grinding (Nespoulet et al., 2008a). This layer is dated to between 105 and 111 ka BP and may be the first recognisable expression of symbolic behaviour in an Aterian context. The likely use of colourants at this site is further suggested by a large dished quartzite nodule (137 × 128 mm) that appears to have traces of a dark red pigment on one of its lateral edges. Other examples of the use of red ochre are present in Layer 21 at Grotte des Pigeons, Taforalt, with associated ages of around 82 ka BP (Bouzouggar et al., 2007) and at Ifri n'Ammar where a flake smeared in red pigment was recovered from layer 32 and dating slightly earlier than 83 ka (Nami and Moser, 2010, Fig. 19). Another interesting feature at El Mnasra, though probably restricted to the upper Aterian layers, is closed fireplaces with edges well defined by limestone slabs as well as open hearths of oval or circular shapes dug into the consolidated clayey sediments (El Hajraoui, 1994). A rudimentary bone industry has also been claimed for the Aterian levels at this site that seems to include awls polished at one end (El Hajraoui, 1994).

Another significant innovation, and one of the most convincing signs of mediated symbolic activity, comes from personal ornaments. These occur in the Aterian predominantly in the form of perforated marine shells of the species *Nassarius gibbosulus* (Fig. 3.3). The use of the sister species *Nassarius circumcintus* and, in one case, of *Columbella*, is also attested. Up to the present, *Nassarius* beads have been recorded at the type site of Oued Djebbana (Vanhaeren et al., 2006) and at four sites with Aterian levels in Morocco (Grotte des Pigeons, Rhafas, Ifri n'Ammar, Contrebandiers) (d'Errico et al., 2009). More recently, they have also been identified at the site of Bizmoune (Bouzouggar and Barton, in press). At Grotte des Pigeons at Taforalt, up until 2009 over 47 beads had been found from stratified deposits in a discrete area towards the rear of the cave. Besides intentional perforation the shells from Grotte des Pigeons, Rhafas, Ifri n'Ammar, and Contrebandiers also share other distinctive features such use-wear around the holes and traces of red ochre, while at Grotte des Pigeons some may have been deliberately heated to change the colour of the bead (d'Errico et al., 2009). So far the examples from Grotte des Pigeons and Ifri n'Ammar have ages in the range of 81–83 ka (Bouzouggar et al., 2007; Richter et al., 2010), with the possibility of earlier occurrences at Contrebandiers (Balter, 2011). At Ifri n'Ammar and Rhafas, *Nassarius* beads occur in layers dated to ca. 80 and 60–70 ka, respectively. It may be significant that at none of the sites mentioned have shell beads been found in any of the oldest Aterian levels. Interestingly, however, *Nassarius* beads have been recovered from Mousterian levels of Skhûl, Israel, that are dated to ca 110 ka and have yielded a number of burials of anatomically modern humans with archaic features (Vanhaeren et al., 2006). This is a theme we shall return to later in this chapter.

Despite the convincing evidence for the use of red pigments and the manufacture of shell beads, two potential indicators of symbolism appear to be missing or invisible in the Aterian. These are the occurrence of abstract and figurative representations of the kind that have been recognised at Blombos Cave, Cape Province, South Africa (Henshilwood et al., 2009), and unequivocal evidence for human burial. Analogies for the latter would be the examples of intact skeletons interred in pits, and sometimes including grave goods, that have been discovered at quasi-contemporary sites in the Near East, such as Skhûl and Qafzeh (Grün et al., 2005; Vandermeersch, 1970). In contrast, human fossils at Aterian sites, where they occur, are only represented by partial skeletal remains and teeth. A case in point is the recently discovered cranium and incomplete skeleton of a seven- to eight-year-child found in deposits dating to 108 ka at Contrebandiers (Balter, 2011). A human phalange and patella have been recovered from layers 27a and 28a at Ifri n'Ammar (Nami and Moser, 2010).

FIGURE 3.3 Red pigment from Aterian layers at El Mnasra, Morocco. *Modified after Nespoulet et al. (2008b).*

3.5. ENVIRONMENTAL CONTEXT FOR DEVELOPMENTS IN THE ATERIAN

As summarised above, the Aterian is known from a wide area of North Africa. Much of this terrain is now covered by Sahara Desert, which today occupies over nine million km^2 and together with the Maghreb and the coastal zones it constitutes the equivalent of about 90% of the surface area of Europe. Although at present only partially habitable, studies of past climatic conditions indicate that the Sahara has been wetter and supported grassland or steppic vegetation at several times in the last 200 ka. In particular a marked humid phase is indicated at the end of MIS 6 when the levels of palaeolakes such as Lake Dakhleh in the Egyptian Western Desert reached a maximum extent of 1700 km^2 (Smith, 2010). Shifts in the monsoon at around this time also permitted increases in seasonal rainfall enough to allow the formation of a series of mega-lakes across the Sahara between 124 and 119 ka (Drake et al., 2008; Osborne et al., 2008; Smith, 2010). It has been proposed that the lakes and interconnecting rivers could have provided a network of routes for human dispersal (Osborne et al., 2008). A later humid period is recorded during MIS 3 between 47 and 42 ka (Giraudi, 2010).

Multiproxy analysis of deep-sea cores is increasing our knowledge of the climatic changes that have affected Aterian populations (Martrat et al., 2004; Sánchez Goñi et al., 2008). These studies, however, have not yet been able to predict the distribution of different biomes and their range change through time. Although published data on the palaeoclimatology and palaeoenvironments of the region are still limited, it seems likely that human presence was not merely confined to periods of increased humidity (Fig. 3.4). Chronological studies show that MP/MSA

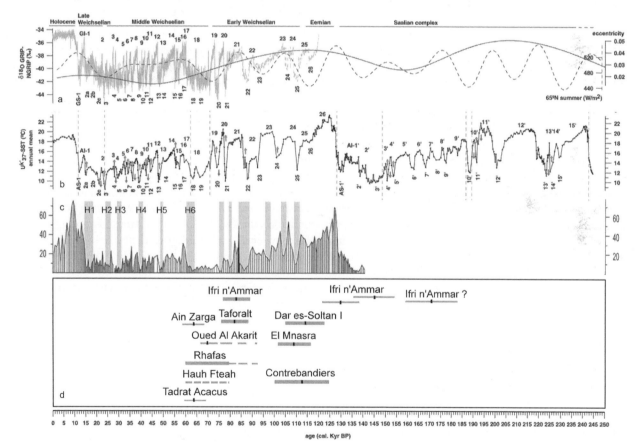

FIGURE 3.4 Paleoenvironmental records of the last two interglacial-to-glacial cycles with a focus on the Mediterranean area compared to available radiometric ages for the Aterian. a) The δ^{18}O in GRIP and NGRIP with the isotopically defined Greenland interstadials (GI) on the top and the subsequent Greenland stadials (GS) on the bottom – over the last climate cycle. The eccentricity of the Earth's orbit and the daily insolation at 65° N during the summer solstice are also shown. b) Sea-surface temperatures in the marine core ODP-977A located in the Alboran Sea. Abrupt changes are defined as warming (red arrows) or cooling (blue arrows) episodes. c) Pollen percentage curve of the Mediterranean forest (deciduous and evergreen *Quercus, Olea, Phillyrea, Pistacia* and *Cistus*) for the last 138 ka in the marine core SU81-18, located on the South-West Iberian margin. d) Radiometric ages for the Aterian with black dots indicating central values and grey bars indicating standard error brackets. Thick grey bars highlight sites/layers with the presence of pigments and/or shell beads. *(a–b: modified after Martrat et al., 2004, c: modified after Sánchez Goñi et al., 2004; d: data from Barton et al., 2009; Schwenninger et al., 2010; Richter et al., 2010; Bouzouggar and Barton, in press)*

"Mousterian" and Aterian populations were present from at least MIS 6 (Smith et al., 2007; Richter et al., 2010) and, following the interglacial, were probably present in the Maghreb, if only at reduced levels, during parts of MIS 4 and the beginning of MIS 3 (Bouzouggar et al., 2007). Evidence for continuous occupation is difficult to demonstrate unequivocally but it is clear that Aterian groups were adapted to a range of habitats from "thermo-Mediterranean" to cooler and more arid environments. At Taforalt, cool, semiarid conditions are indicated in levels with *Nassarius* shell beads by the presence of *Ctenodactylus* spp. (gundi), which now occupy areas well to the south in steppe and rock outcrops along the margins of the Sahara. The existence of equids and ostrich in these levels also implies more open steppic conditions (Bouzouggar et al., 2007) during this period of occupation. Similarities in the microfaunal indicators have been recognised in the Aterian OS layers (*Occupation Supérieure*) at Ifri n'Ammar dating to greater than 83 ka (Hutterer in Nami and Moser, 2010).

3.6. DISCUSSION

From this brief overview of the Aterian a number of important discussion points emerge that have wider implications for understanding the MP/MSA of North Africa, and for making more meaningful comparisons at a transcontinental level. The Aterian appears in many respects as a fascinating and unique case in that it challenges current theories and expectations about the origins of cultural modernity and, by the same token, provides clues that allow clearer insight into the process that has led our species to create human cultures as we know them. For some scholars (McBrearty and Brooks, 2000; Marean et al., 2007; Mellars, 2005; Klein, 2000) the emergence of cultural modernity is, and can only be, the direct outcome of biological factors that have produced our species in Africa. In such a theoretical frame African anatomically modern human populations are expected to behave in a "modern way" and do so little by little according to gradualistic models (McBrearty and Brooks, 2000; Marean et al., 2007) or as a result of abrupt changes in case of "revolutionary" models (Klein, 2000; Mellars, 2006). Whatever scenario is favoured, cultural modernity is primarily reflected and inextricably linked to the identification of sometimes rapid cultural turnovers related to changes in the geographic expressions of cultural entities (Klein, 2009; McBrearty and Brooks, 2000; Henshilwood and Marean, 2003; Shennan, 2001; Powell et al., 2009). In many instances, such changes are observably synchronic and closely linked to environmental changes, and coincident with more or less abrupt ecological clines (Lahr and Foley, 2004; Banks et al., 2008). There are also manifest implications for relating such changes to population demography and migration patterns (Henrich, 2004). Seen from this perspective the Aterian appears at a first sight to present an unlikely paradox.

In particular, on the basis of recently obtained ages this technocomplex may have covered a time span of more than 70,000 years, successfully traversing major environmental changes, and extending across a territory of 1,000,000 km^2. The bulk of newly dated Aterian sites is contemporaneous with the period of variable climate in the earlier phases of MIS 5 (MIS 5e-c), and the beginning of MIS 5a, which is characterised by a gradual decrease of temperatures and precipitation. It perhaps also persisted into the cool, dry conditions of MIS 4, and continued until the onset of MIS 3. This apparently occurred without generating archaeologically noticeable changes in the fundamental attributes of the Aterian technical system. It is therefore possible to comprehend, in the light of the above assumptions, why the features that made the Aterian so successful can at the same time also be perceived as symptomatic of its inherent limitations.

In other contexts, the Stillbay and the Howieson Poort in Southern Africa and, later on, Early Upper Palaeolithic archaeological cultures in Europe and the Near East such as the Châtelperronian, the Uluzzian, the Szeletian, the Ahmarian and the proto-Aurignacian can all be seen as more appropriate examples of the rapid cultural change model that we associate with "modernity". This is emphasised by the fact that innovations appear in some of these relatively short cultural episodes such as the Stillbay as a package and invoke from the start different technological (bone technology, pyrotechnology and sophisticated knapping) and symbolic (abstract engraving and bead use) domains (Backwell et al., 2008; Henshilwood et al., 2009; d'Errico et al., 2009; Mourre et al., 2010). Aterian innovations, in contrast, stem from an old root and do not seem to arise synchronously or stimulate detectable changes in the remainder of the cultural system. The disappearance of the Stillbay at the onset of the climatic deterioration of the MIS 4 suggests that, though remarkable in many respects, such a novel integration of complex technical and symbolic behaviours was a fragile cultural experiment easily weakened by climatic adversity. If an age of 60 ka or younger is confirmed for late Aterian assemblages, this would imply that this cultural adaptation in contrast to South Africa was able to accommodate the marked environmental changes of MIS 4 without fatal interruptions in cultural transmission.

A major challenge of current research on the Aterian is to fully document symbolically mediated behaviours associated with this technocomplex, identify the time span during which they were at work and put forward a viable explanation for the resulting pattern. No personal ornaments are attested for the moment at later i.e., post-MIS 4 Aterian sites. If this is confirmed by future research, we will need to explain what mechanisms have produced the loss of some innovations while preserving the core of the technical system and what implications this may have for comparative examples in other parts of Africa. Strictly speaking, none of the mechanisms that have been proposed so far to account for the spread, maintenance and loss of innovations appears to best fit the Aterian trajectory. According to Henrich (2004) and Powell et al. (2009), demographic growth triggered by climatic change would facilitate the spread and consolidation of new technologies allowing more effective subsistence strategies. Development of symbolically mediated behaviours reflects in this model the need for bounding practices supporting stable social networks. However, no clear indicators of population growth and technological change are seen in the Aterian in relation and in parallel with the production of personal ornamentations apart perhaps from the marginal presence of small bifacial points in combination with pedunculates. Habitat contraction created by rapid rise in sea level associated with glacial terminations and connected shift in vegetation biomes are seen by Compton (2011) as the main triggering mechanisms for the development of modern cultures in South Africa. Rapid flooding of exposed areas of the coastal plain of southern cape would have increased population density and competition while, at the same time, isolating some subpopulations in areas constrained by geographic barriers and ecological changes. This process would have selected for populations who expanded their diet to include marine resources and large hunted animals. In North Africa, sea-level changes did not significantly affect coastline delineation, nor was the hinterland low in resources. So, rather than isolating populations, the initial phases of each interglacial opened up pathways through the Sahara, which paradoxically reduced the isolation of North Africa. Contrary to the predictions of Compton's model, innovations such as personal ornamentation appear in the Aterian at the end of the last interglacial and disappear at the very moment in which they should theoretically have emerged, i.e., with the beginning of the MIS 4 climatic deterioration (Fig. 3.4). At present, we do not have any wide-ranging explanation to account for this phenomenon apart from the idea proposed in d'Errico et al. (2009) that MIS 4 apparently broke exchange networks between inland and coastal sites required for the maintenance of bead traditions which, for some reason, left the technical and economic systems unaffected. The recent identification at Ifri n'Ammar of an interstratification of Mousterian and Aterian layers apparently spanning many millennia (Richter et al., 2010) adds a supplementary level of complexity to the interpretation of this phenomenon. If correct, and setting aside the unlikely hypothesis of disappearance and subsequent reinvention of the same technology, it suggests that the example of Ifri n'Ammar reflects a long-term fluctuating boundary between slightly different cultural adaptations. It follows that because of either historical contingencies or, more probably, shifts in habitat triggered by climate change, Mousterian and Aterian have represented equally viable adaptive solutions for a long time in this region. In contrast, it appears that the tradition of using *Nassarius* shells for bead making first appeared in the Levant in the Mousterian at Skhûl Cave. However, this does not imply cultural transmission from the east end of the Mediterranean to the west. Indeed, it may be time to substitute a different term for the Mousterian in the Maghreb (Kleindienst, 1998).

Taxonomic affiliation of the makers of the Aterian enhances the apparent mismatch associated with these two technocomplexes. Human remains found at Aterian sites are interpreted as clearly belonging to members of our own species, although displaying some archaic features. In the light of the above, such an association may appear to contradict the often expressed and inescapable link between cultural modernity and the biological process that has led to our present-day anatomy and associated genetic heritage (Marean et al., 2007; Hublin, 2002). This is even more exaggerated when considering that Mousterian-like assemblages found in the same region are also associated with modern human remains and display some commonalities with the Levantine Mousterian, made by both moderns and Neanderthal populations. If the people associated with the Levantine Mousterian made personal ornaments, and if the Levantine Mousterian displays broad commonalities with the African Mousterian, how is it that the former also includes personal ornaments and the latter does not? Part of the answer to this question may lie in our inadequate understanding of the industries termed "Mousterian" in the Maghreb which, as we have suggested above, may be in need of significant revision.

The above idiosyncrasies in relation to the Aterian only disappear when we think of the emergence of cultural innovations in general and symbolically mediated behaviours in particular as driven by a complex nonlinear process involving cognitive pre-adaptations, the nature of regional cultural traditions, ecological conditions and their impact on demography. In North Africa, symbolic behaviour and the use of shell beads in the Aterian do not appear as the obligatory outcome of a special evolutionary pathway but rather as a widespread, relatively short-lived, cultural experiment stemming from a long-term successful cultural adaptation. Although the precise mechanisms that have produced this intriguing phenomenon largely remain to be discovered, the production of a symbolic material culture within a cultural system often considered in the past as stuck in unmodifiable cultural traditions and incapable of such a performance reveals the plastic nature of MSA/MP adaptations and their ability to cross in their own way the Rubicon towards modernity well before the Upper Palaeolithic.

ACKNOWLEDGEMENTS

We would like to thank Mathieu Leroyer for the artefact illustrations in Fig. 3.1 and Ian Cartwright for his help in producing this figure. Roland Nespoulet is also acknowledged for giving permission to reproduce Fig. 3.3. Francesco d'Errico acknowledges financial support from the European Research Council (FP7/2007/2013;/ERC Grant TRACSYMBOLS n°249587) and the Programme PROTEA of the French Ministry of Higher Education and Research. Nick Barton would like to thank The British Academy (LRG 39859) and the Natural Environment Research Council (EFCHED project and RESET consortium grant) for supporting research presented in this paper.

REFERENCES

Alimen, H., 1957. The Prehistory of Africa. Translated by A.H. Brodrick. Hutchinson, London.

Antoine, M., 1938. Notes de préhistoire marocaine: XIV – Un cône de résurgence du Paléolithique moyen à Tit Mellil, près de Casablanca. Bulletin de la Société Préhistorique du Maroc 12, 3–95.

Backwell, L.R., d'Errico, F., Wadley, L., 2008. Middle Stone Age bone tools from the Howiesons Poort layers, Sibudu Cave, South Africa. J. Archaeol. Sci. 35 (6), 1566–1580.

Balout, L., 1965. Données nouvelles sur le problème du moustérien en Afrique du Nord. Actas del V Congresso Panafricano de Prehistoria y de Estudio del Cuaternario. Publicaciones del Museo Arqueologico Sant Cruz de Tenerife, pp. 137–145.

Balter, M., 2011. Was North Africa the launch pad for Modern Human Migrations? Science 331, 20–23.

Banks, W.E., D'Errico, F., Townsend Peterson, A., Kageyama, M., Sima, A., Sánchez Goñi, M.-F., 2008. Neanderthal extinction by competitive exclusion. Plos One 3 (12), 1–8.

Barich, B.E., Garcea, E.A.A., 2008. Ecological patterns in the Upper Pleistocene and Holocene in the Jebel Gharbi, Northern Libya: chronology, climate and human occupation. Afr. Archaeol. Rev. 25, 87–97.

Barker, G., Antoniadou, A., Armitage, S., Brooks, I., Candy, I., Connell, K., Douka, K., Drake, N., Farr, L., Hill, E., Hunt, C., Inglis, R., Jones, S., Lane, C., Luccharini, G., Meneely, J., Morales, J., Mutri, G., Prendergast, A., Rabett, R., Reade, H., Reynolds, T., Russell, N., Simpson, D., Smith, B., Stimpson, C., Twati, M., White, K., 2010. The cyrenaican prehistory project 2010: the fourth season of investigations of the Haua Fteah cave and its landscape, and further results from the 2007–2009 fieldwork. Libyan Studies 41, 63–88.

Barton, R.N.E., Bouzouggar, A., Collcutt, S., Schwenninger, J.-L., Clark-Balzan, L., 2009. OSL dating of the Aterian levels at Dar es-Soltan I (Rabat, Morocco) and implications for the dispersal of modern Homo sapiens. Quat. Sci. Rev. 28, 1914–1931.

Bouzouggar, A., 1997. Economie des matières premières et du débitage dans la séquence atérienne de la grotte d'El Mnasra I (ancienne grotte des Contrebandiers-Maroc). Préhist. Anthrop. Médit. 6, 35–52.

Bouzouggar, A., Barton, N., Vanhaeren, M., d'Errico, F., Collcutt, S., Higham, T., Hodge, E., Parfitt, S., Rhodes, E., Schwenninger, J.-L., Stringer, C., Turner, E., Ward, S., Moutmir, A., Stambouli, A., 2007. 82,000-year-old shell beads from North Africa and implications for the origins of modern human behavior. Proc. Natl. Acad. Sci. U S A 104, 9964–9969.

Bouzouggar, A., Barton, R.N.E., 2012. The identity and timing of the Aterian in Morocco. In: Hublin, J.-J., McPherron, S. (Eds), Modern Origins: A North African Perspective. Springer: New York, 93–106.

Bouzouggar, A., Kozlowski, J.K., Otte, M., 2002. Etude des ensembles lithiques atériens de la grotte d'El Aliya à Tanger (Maroc). L'Anthropologie 106, 207–248.

Byrne, R., 1995. The Thinking Ape. Oxford University Press, Oxford.

Camps, G., 1974. Les civilisations préhistoriques de L'Afrique du Nord et du Sahara. Doin, Paris.

Carrière, G., 1886. Quelques stations préhistoriques de la Province d'Oran. Bull. Soc. Géog. Archéol. Oran VI, 136–154.

Caton-Thompson, G., 1946. The Aterian industry: its place and significance in the Palaeolithic world. J. R. Anthropol. Inst. G.B. Irel. 76 (2), 87–130.

Clark, J.D., 1982. The cultures of the Middle Palaeolithic/Middle Stone Age. In: Clark, J.D. (Ed.), From the Earliest Times to c. 500 BC. The Cambridge History of Africa, Vol. 1. Cambridge University Press, Cambridge, pp. 248–341.

Clark, J.D., 1993. African and Asian perspectives on the origins of Modern Humans. In: Aitken, M.J., Stringer, C.B., Mellars, P.A. (Eds.), The Origin of Modern Humans and the Impact of Chronometric Dating. Princeton University Press, Princeton, pp. 148–178.

Compton, J.S., 2011. Pleistocene sea-level fluctuations and human evolution on the southern coastal plain of South Africa. Quaternary Science Reviews 30 (5–6), 506–527.

d'Errico, F., Stringer, C., 2011. Evolution, Revolution or Saltation scenario for the emergence of modern cultures? Philos. Trans. R. Soc. Lond. B Biol. Sci. 366, 1060–1069.

d'Errico, F., Vanhaeren, M., Barton, N., Bouzouggar, J., Mienis, H., Richter, D., Hublin, J.J., McPherron, S.P., Lozouet, P., 2009. Additional evidence on the use of personal ornaments in the Middle Paleolithic of North Africa. Proc. Natl. Acad. Sci. U S A 106 (38), 16051–16056.

Deacon, T., 1997. The Symbolic Species. W.W. Norton & Co, New York.

Debénath, A., Raynal, J.-P., Roche, J., Texier, J.-P., Ferembach, D., 1986. Stratigraphie, habitat, typologie et devenir de l'Atérien Marocain: données récentes. L'Anthropologie 90, 233–246.

Donald, M., 1991. Origins of the Modern Mind. Harvard University Press, Cambridge, MA.

Drake, N.A., El-Hawat, A.S., Turner, P., Armitage, S.J., Salem, M.J., White, K.H., McLaren, S., 2008. Palaeohydrology of the Fazzan Basin and surrounding regions: the last 7 million years. Palaeogeogr. Palaeoclimatol. Palaeoecol. 263, 131–145.

El Hajraoui, M.A., 1994. L'industrie osseuse atérienne de la grotte d'El Mnasra (Région de Témara, Maroc). LAPMO Université de Provence CNRS. Préhist. Anthrop. Médit. 3, 91–94.

Garcea, E.A.A. (Ed.), 2001. Uan Tabu in the Settlement History of the Libyan Sahara. All'Insegna del Giglio, Firenze.

Garcea, E.A.A., 2009. The evolutions and revolutions of the Late Middle Stone Age and Lower Later Stone Age in north-west Africa. In: Camps, M., Szmidt, C. (Eds.), The Mediterranean from 50,000 to 25,000 BP: Turning Points and New Directions. Oxbow Books, Oxford, pp. 49–64.

Garcea, E.A.A., 2010. The spread of Aterian peoples in North Africa. In: Garcea, E.A.A. (Ed.), South-Eastern Mediterranean Peoples Between 130,000 and 10,000 Years Ago. Oxbow Books, Oxford, pp. 37–53.

Giraudi, C., 2010. Geology and the Palaeoenvironment. In: Barich, B.E., Garcea, E.A.A., Giraudi, C., Lucarini, G., Mutri, G. (Eds.), The Latest Research in the Jebel Gharbi (Northern Libya): Environment and Cultures from MSA to LSA and the First Neolithic Findings. Libya Antiqua n.s. 5, pp. 237–240.

Grün, R., Stringer, C.B., McDermott, F., Nathan, R., Porat, N., Robertson, S., Taylor, L., Mortimer, G., Eggins, S., McCulloch, M., 2005. U-series and ESR analyses of bones and teeth relating to the human burials from Skhul. J. Hum. Evol. 49, 316–334.

Harvati, K., Hublin, J.J., 2012. Morphological continuity of the face in the late middle and upper Pleistocene hominins from Northwestern Africa – A 3D geometric morphometric analysis In: Hublin, J.J., McPherron, S. (Eds), Modern Origins: A North African Perspective. Springer: New York, 179–188.

Henrich, J., 2004. Demography and cultural evolution: why adaptive cultural processes produced maladaptive losses in Tasmania. Am. Antiq. 69 (2), 197–214.

Henshilwood, C.S., d'Errico, F., Watts, I., 2009. Engraved ochres from the Middle Stone Age levels at blombos cave, South Africa. J. Hum. Evol. 57, 27–47.

Henshilwood, C., Marean, C.W., 2003. The origin of modern human behavior: critique of the models and their test implications. Curr. Anthropol. 44 (5), 627–649.

Hublin, J.J., 2000. Modern-nonmodern hominid interactions: a mediterranean perspective. In: Bar-Yosef, O., Pilbeam, D. (Eds), The Geography of Neandertals and Modern Humans in Europe and the Greater Mediterranean. Peabody Museum Bulletin 8. Cambridge, pp. 157–182.

Hublin, J.-J., 2002. Northwestern African Middle Pleistocene hominids and their bearing on the emergence of Homo sapiens. In: Barham, L., Robson-Brown, K. (Eds.), Human Roots: Africa and Asia in the Middle Pleistocene. Western Academic & Specialist Press, Bristol, pp. 99–121.

Hublin, J.-J., Tillier, A.-M., Tixier, J., 1987. L'humérus d'enfant moustérien (Homo 4) du jebel irhoud (maroc) dans son contexte archéologique. Bulletin Et Mémoires De La Société d'Anthropologie De Paris 4 (série XIV), 115–142.

Hublin, J.J., Verna, C., Bailey, S., Smith, T., Olejniczak, A., Sbihi-Alaoui, F.Z., Zouak, M., 2012. Dental evidence from the aterian human populations of Morocco. In: Hublin, J.J., McPherron, S. (Eds), Modern Origins: A North African Perspective. Springer: New York, 189–204.

Klein, R.G., 2000. Archaeology and the evolution of human behavior. Evol. Anthropol. 17–35.

Klein, R.G., 2009. The Human Career: Human Biological and Cultural Origins. The University of Chicago Press, Chicago.

Kleindienst, M.R., 1998. What is the Aterian? The view from Dakhleh Oasis and the Western Desert, Egypt. In: Marlow, M., Mills, A.J. (Eds.), The Oasis Paper 1: The Proceedings of the First Conference of the Dakhleh Oasis Project. Oxbow Books, Oxford, pp. 1–14.

Kleindienst, M.R., Smith, J.R., Adelsberger, K.A., 2009. The Kharga Oasis Prehistory Project (KOPP), 2008 field season: part 1. Geoarchaeology and Pleistocene Prehistory. Nyame Akuma 71, 18–30.

Lahr, M., Foley, R., 1994. Multiple dispersals and modern human origins. Evol. Anthropol. 3 (2), 48–60.

Marean, C.W., Bar-Matthews, M., Bernatchez, J., Fisher, J., Goldberg, P., Herries, A., Jacobs, Z., Jerardino, A., Karkanas, P., Minichillo, T., Nilssen, P.J., Thompson, E., Watts, I., Williams, H.M., 2007. Early human use of marine resources and pigment in South Africa during the Middle Pleistocene. Nature 449, 905–908.

Marean, C.W., 2010. Pinnacle point cave 13B (Western Cape Province, South Africa) in context: the Cape floral kingdom, shellfish, and modern human origins. J. Hum. Evol. 59, 425–443.

Martrat, B., Grimalt, J.O., Lopez-Martinez, C., Cacho, I.C., Sierro, F.J., Flores, J.A., Zahn, R., Canals, M., Curtis, J.H., Hodell, D.A., 2004. 4 Abrupt temperature changes in the western Mediterranean over the past 250,000 years. Science 306, 1762–1765.

McBrearty, S., Brooks, A.S., 2000. The revolution that wasn't: a new interpretation of the origin of modern human behavior. J. Hum. Evol. 39, 453–563.

Mellars, P., 2005. The impossible coincidence. A single-species model for the origins of modern human behaviour in Europe. Evol. Anthropol. 14, 12–27.

Mellars, P., 2006. Why did modern human populations disperse from Africa ca. 60,000 years ago? A new model. Proc. Natl. Acad. Sci. U S A 103 (5), 9381–9386.

Morel, J., 1974a. Nouvelles datations absolues de formations littorales et de gisements préhistoriques de l'Est algérien. Bulletin De La Société Préhistorique Française 71, 103–105.

Morel, J., 1974b. La station éponyme de l'Oued Djebbana à Bir-el-Ater (Est algérien), contribution à la connaissance de son industrie et de sa faune. L'Anthropologie 78, 53–80.

Mourre, V., Villa, P., Henshilwood, C.S., 2010. Early use of pressure flaking on lithic artifacts at blombos cave, South Africa. Science 330 (6004), 659–662.

Nami, M., Moser, J., 2010. La Grotte D'Ifri n'Ammar. Tome 2: Le Paléolithique Moyen. Reichert Verlag, Wiesbaden.

Nespoulet, R., Debénath, A., El Hajraoui, M.A., Michel, P., Campas, E., Oujaa, A., Ben-Ncer, A., Lacombe, P., Amani, F., Stoetzel, E., Boudad, L. 2008b. Le contexte archéologique des restes humains atériens de la région de Rabat-Témara (Maroc): Apport des fouilles des grottes d'El Mnasra et d'El Harhoura 2. Actes RQM4, Oujda, 2008, 356–375.

Nespoulet, R., El Hajraoui, M.A., Amani, F., Ben-Ncer, A., Debénath, A., El Idrissi, A., 2008a. Palaeolithic and Neolithic occupations in the Témara region (Rabat, Morocco): recent data on hominin contexts and behaviour. Afr. Archaeol. Rev. 25, 21–40.

Osborne, A.H., Vance, D., Rohling, E.J., Barton, N., Rogerson, M., Fello, N., 2008. A humid corridor across the Sahara for the migration of early modern humans out of Africa 120,000 years ago. Proc. Natl. Acad. Sci. U S A 105, 16444–16447.

Powell, A., Shennan, S., Thomas, M.G., 2009. Late Pleistocene demography and appearance of modern behaviour. Science 324, 1288–1301.

Reygasse, M., 1921–1922. Études de Paléthnologie Maghrébine (deuxième série). Recueil des Notices et Mémoires de la Société Archéologique, Historique et Géographique du Constantine LIII, 159–204.

Reygasse, M., 1922. Note au sujet de deux civilizations préhistoriques africaines pour lesquelles deux termes nouveaux me paraissent devoir être employés. XLVIè Congrès De L'association Française Pour L'avancement Des Sciences, Montpellier, pp. 467–472.

Richter, D., Moser, J., Nami, M., Eiwanger, J., Mikdad, A., 2010. New chronometric data from Ifri n'Ammar (Morocco) and the chronostratigraphy of the Middle Palaeolithic in the Western Maghreb. J. Hum. Evol. 59 (6), 672–679.

Roset, J.-P., Harbi-Riahi, M., 2007. El Akarit: Un Site Archéologique Du Paléolithique Moyen Dans Le Sud De La Tunisie. Editions Recherches sur les Civilisations, Paris.

Rossano, M.J., 2010. Making friends, making tools, and making symbols. Curr. Anthropol. 51, 89–98.

Rumbaugh, D.M., Washburn, D.A., 2003. The Intelligence of Apes and Other Rational Beings. Yale University Press, New Haven, CT.

Sánchez Goñi, M.F., Landais, A., Fletcher, W.J., Naughton, F., Desprat, S., Duprat, J., 2008. Contrasting impacts of Dansgaard–Oeschger events over a western European latitudinal transect modulated by orbital parameters. Quaternary Science Reviews 27, 1136–1151.

Schwenninger, J.-L., Collcutt, S.N., Barton, N., Bouzouggar, A., Clark-Balzan, L., El Hajraoui, M.A., Nespoulet, R., Debénath, A., 2010. A new luminescence chronology for Aterian cave sites on the Atlantic coast of Morocco. In: Garcea, E.A.A. (Ed.), South-Eastern Mediterranean Peoples Between 130,000 and 10,000 Years Ago. Oxbow Books, Oxford, pp. 18–36.

Shea, J.J., 2011. *Homo sapiens* is as *Homo sapiens* was. Behavioral variability versus "behavioral modernity" in Paleolithic Archaeology. Curr. Anthropol. 52 (1), 1–34.

Shennan, S., 2001. Demography and cultural innovation: a model and its implications for the emergence of modern human culture. Camb. Archaeol. J. 11, 5–16.

Smith, J., Giegengack, R., Schwarcz, H.P., McDonald, M.A.A., Kleindienst, M.R., Hawkins, A.L., Churcher, C.S., 2004. Reconstructing Pleistocene pluvial environments and occupation through the stratigraphy and geochronology of fossilspring tufas, Kharga Oasis, Egypt. Geoarchaeology 19, 407–439.

Smith, J.R., 2010. Palaeoenvironments of eastern North Africa and the Levant in the late Pleistocene. In: Garcea, E.A.A. (Ed.), South-Eastern Mediterranean Peoples Between 130,000 and 10,000 Years Ago. Oxbow Books, Oxford, pp. 6–17.

Smith, T.M., Tafforeau, P., Reid, D.J., Grün, R., Eggins, S., Boutakiout, M., Hublin, J.-J., 2007. Earliest evidence of modern human life history in North African early *Homo sapiens*. Proc. Natl. Acad. Sci. U S A 104, 6128–6133.

Tixier, J., 1958–1959. Les pièces pédonculés de l'Atérien. Libyca VI–VII, 127–158.

Tixier, J., 1967. Procédés d'analyse et questions de terminologie concernant l'étude des ensembles industriels du Paléolithique récent et de l'Epipaléolithique dans l'Afrique du nord-ouest. In: Bishop, W.W., Clark, J.D. (Eds.), Background to Evolution in Africa. University of Chicago Press, Chiocago, pp. 771–820.

Tomasello, M., Carpenter, M., Call, J., Behne, T., Moll, H., 2005. Understanding and sharing intentions: the origins of cultural cognition. Behav. Brain Sci. 28, 675–735.

Van Peer, P., 1991. Interassemblage variability and Levallois styles: the case of the North African middle Palaeolithic. J. Anthropol. Archaeol. 10, 107–151.

Vandermeersch, B., 1970. Une sépulture moustérienne avec offrandes decouverte dans la grotte de Qafzeh. Comptes Rend. Acad. Sci. 268, 298–301.

Vanhaeren, M., d'Errico, F., Stringer, C.B., James, S.L., Todd, J.A., Mienis, H.K., 2006. Middle Paleolithic shell beads in Israel and Algeria. Science 312, 1785–1788.

Wendorf, F., Schild, R., 1976. The Prehistory of the Nile Valley. Academic Press, New York.

Wengler, L., 1997. La transition du Moustérien à l'Atérien. L'Anthropologie 101 (3), 448–481.

Whiten, A., 2005. The second inheritance systems of chimpanzees and humans. Nature 437, 52–55.

Personal Ornaments and Symbolism Among the Neanderthals

João Zilhão

ICREA Research Professor at the University of Barcelona, (Seminari d'Estudis i Recerques Prehistòriques; Departament de Prehistòria, Història Antiga i Arqueologia; Facultat de Geografia i Història; C/ Montalegre 6; 08001 Barcelona; Spain), joao.zilhao@ub.edu

"The distinctive faculties of Man are visibly expressed in his elevated cranial dome—a feature which, though much debased in certain savage races, essentially characterises the human species. But, considering that the Neanderthal skull is eminently simial, both in its general and particular characters, I feel myself constrained to believe that the thoughts and desires which once dwelt within it never soared beyond those of a brute. The Andamaner, it is indisputable, possesses but the dimmest conceptions of the existence of the Creator of the Universe: his ideas on this subject, and on his own moral obligations, place him very little above animals of marked sagacity; nevertheless, viewed in connection with the strictly human conformation of his cranium, they are such as to specifically identify him with Homo sapiens. *Psychical endowments of a lower grade than those characterising the Andamaner cannot be conceived to exist: they stand next to brute benightedness. (…) Applying the above argument to the Neanderthal skull, and considering … that it more closely conforms to the brain-case of the Chimpanzee, … there seems no reason to believe otherwise than that similar darkness characterised the being to which the fossil belonged"* (King, 1864; pp. 96).

4.1. INTRODUCTION

As William King's seminal definition well illustrates, cognitive impairment has been central to the notion that Neanderthals were a different species from the very beginning of human evolution studies. King's paper is the written version of a presentation to the 1863 Newcastle meeting of the British Association for the Advancement of Science, and includes a footnote where, explaining that his mind had since changed, he went even further: "I now feel strongly inclined to believe that it [*Homo Neanderthalensis*] is not only specifically but generically distinct from Man." This mid-19th-century perception of Neanderthal-ness was strengthened by Boule's (1913) assessment of the complete skeleton discovered in 1908 at La Chapelle-aux-Saints, in France, as belonging to an ape-like, hunchback creature with a stooping, imperfectly bipedal gait. The graphic reconstruction of the creature's life appearance, directly inspired by Boule, powerfully conveyed the brute benightedness to which King's Andamaner stood next and in which his Neanderthal would have been squarely immersed (Fig. 4.1).

It is my contention here that these early views continue to condition thinking about the Neanderthals, and this among the general public as much as within academia. Only such a deep paradigmatic bias can explain the widespread application of double standards in the evaluation of the empirical evidence that underpins ongoing debates about the place of Neanderthals in human evolution and the nature of their relationship with extant humanity. I have made this argument before (Zilhão, 2001, 2011), as have others (e.g., Trinkaus and Shipman, 1994; Wolpoff and Caspari, 1996; Roebroeks and Corbey, 2001; Speth, 2004), but research developments of the past decade have intensely illuminated the extent to which this remains a major problem. Therefore, it is useful to come back to the issue to further hammer in the point.

The Recent African Origin (RAO) or Mitochondrial Eve model of modern human origins that dominated the field since the mid-1980s proposed that all extant humans descended from a small East African population that speciated into *H. sapiens*. Thanks to the competitive advantage granted by the biological changes embodying the transformation, and attendant behavioural consequences, those first fully human beings would have rapidly expanded from their source area into adjacent regions of Africa, first, and then into Eurasia (e.g., Stringer and Andrews, 1988; Stringer and Gamble, 1993). Along the way, the local "archaic" populations encountered would have been replaced, extinction without descent having thus been the fate of such aboriginal groups, namely the Neanderthals.

FIGURE 4.1 Reconstruction of the La Chapelle-aux-Saints Neanderthal by the Czech artist František Kupka. Produced in 1909, this reconstruction was heavily influenced by Marcelin Boule's view of the skeleton as belonging to a creature not fully human.

Against the background of its present-day primate relatives, advanced cognition is the hallmark of the human species. Therefore, the underlying assumption of such RAO views was that the material evidence for a species-level distinction of "moderns" (in skeletal morphology as much as in material culture) also served as a proxy for a species-level distinction at the cognitive level – essentially the same point that King had made. Although differences of detail existed, the notion was, basically, that people who were like "us" anatomically should also have been like "us" cognitively. In other words, they would have been endowed with symbolic thinking and language. Conversely, people who were not like "us" anatomically could not have been like "us" cognitively either. Thus, Neanderthals and other coeval archaic forms of humanity living elsewhere in Africa, Europe or Asia were seen as somehow handicapped by comparison, lacking in symbolic thinking and language, or having only primitive, inferior versions of them (e.g., Davidson and Noble, 1996).

These notions were put into practice through the use of definitions containing explicit criteria upon which "behavioural modernity" could be empirically recognised. Initially, such criteria were based on the archaeological record of Europe, and essentially consisted of lists of traits separating the Middle from the Upper Palaeolithic (e.g., White, 1982). The inadequacy of such lists gradually became apparent as the realisation sank in that the emergence of anatomical modernity in Africa went back beyond 100,000 years ago and, therefore, predated the Upper Palaeolithic of Europe by more than fifty millennia. This carried the implication that either the emergence of anatomical and behavioural/cognitive modernity had to be dissociated – as in Klein's (2003) view that the capacity for language resulted from a genetic mutation occurring among African moderns no earlier than some 50,000 years ago – or new definitions were needed. Eventually, this latter alternative prevailed and, over the last decade, building on McBrearty and Brooks' (2000) vast survey of the evidence, archaeologists working in Africa developed a set of criteria adapted to the nature of their record and based on features whose presence/absence served to assess whether the species-specific "modern behaviour" of "modern humans" was or was not reflected in the late Middle and early Upper Pleistocene sites of that continent.

The following are two summary statements concerning indicators of "behavioural modernity" in the archaeological record of Africa whose validity is widely accepted by palaeoanthropologists:

"Artifacts or features carrying a clear, exosomatic symbolic message, such as personal ornaments, depictions, or even a tool clearly made to identify its maker" (Henshilwood and Marean, 2003).

"Complex use of technology, namely the controlled use of fire as an engineering tool to alter raw-materials; for example, heat pre-treating poor-quality siliceous rocks to enhance their flaking properties" (Brown et al., 2009).

Overall, there is little question that these and other authors (e.g., d'Errico et al., 2003) did a very good job in highlighting the extent to which, using such criteria, one could identify the crossing of a significant behavioural threshold in the archaeological record of the African continent, and especially so in that of southern Africa, sometime after 150,000 years ago. This recognition satisfied the expectations of the RAO model in that it supported the notion that anatomical and behavioural modernity emerged in tandem, as one would expect if both resulted from the differentiation of a new species. Put another way, in any particular time/space configuration, the absence of those defining features from the record would reflect the fact that modern humans were not yet in existence, while their presence would reflect the time of emergence (at the source) or immigration (elsewhere) of the new species.

The above was evidently *necessary* for the RAO model to be supported, but it was not *sufficient*. What was rarely, if ever, asked from within this view of modern human origins was: Does the application of such criteria also enable us to categorise the Neanderthals and other anatomically archaic peoples as behaviourally nonmodern, i.e., as behaving like Africans of the period before 150,000 years ago even when living no more than just 100,000 or 50,000 years ago? Only if the answer to this question was "yes" would the model be logically consistent and, indeed, such a positive answer has been widely asserted or assumed, often on the basis of exceedingly cursory reviews of the Eurasian evidence that tended to explain away counter examples on the basis of the evidence being insecure (e.g., as coming from "old excavations"). In retrospect, however, we have to ask ourselves the following question: As RAO and the subsequent African-based "behavioural modernity" or "human revolution" paradigms were being developed, what knowledge was already in existence concerning the behaviour of the Neanderthals? Is it true that all secure evidence then available suggested that European peoples of the first part of the last glacial behaved like the Africans of previous glacials?

4.2. "NEANDERTHAL BEHAVIOUR": THE MID-20TH CENTURY KNOWLEDGE BASE AND THE HUMAN REVOLUTION

The above-quoted African-based definitions of behavioural modernity put forth by Henshilwood and Marean (2003) and by Brown et al. (2009) emphasise symbolic artefacts or features as well as complex technology. Concerning symbolism, let us first take a look at the evidence that has been available since the 1930s' publication of the excavation of the French rockshelter of La Ferrassie, in the Dordogne (Peyrony, 1934; Defleur, 1993), where a number of individual burials containing the remains of seven people deceased in infancy, childhood and adult age were found in a single level of a deeply stratified deposit. The Neanderthal morphology of the remains is unquestionable, the integrity of the context is documented by the very fact of skeletal articulation and the associated stone tools have been used by archaeologists for decades to define a specific variant of the Eurasian Middle Palaeolithic, the Ferrassie-type Mousterian (Bordes, 1968). The chronology of this culture elsewhere in France indicates that the cemetery use of La Ferrassie by Neanderthals dates between 60,000 and 75,000 years ago. All of this is well known and rather uncontroversial, but that is not so with two other significant features of these burials. The first is that a bone fragment decorated with four sets of parallel incisions lay alongside the La Ferrassie 1 individual, an adult male; the second is that the La Ferrassie 6 individual, a 3- to 5-year-old child, had been interred in a deep pit covered by a limestone slab whose inferior face had been decorated with cup holes (Fig. 4.2).

A second example of symbolism among Eurasian Neanderthals concerns the Châtelperronian levels of the Grotte du Renne, at Arcy-sur-Cure, France (Leroi-Gourhan, 1961; Leroi-Gourhan and Leroi-Gourhan, 1965; d'Errico et al., 1998; David et al., 2001; Schmider, 2002; Zilhão, 2006, 2007, 2011; Caron et al., 2011). Although the fieldwork, directed by A. Leroi-Gourhan, took place more than 50 years ago (1949–1963), this was probably the first site in the world to be excavated with modern techniques (stratigraphic excavation and area exposure of occupation surfaces, spatial plotting of key finds and features and systematic sieving of the deposits) over an extended period

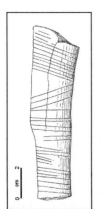  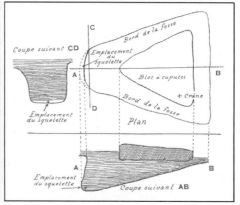

FIGURE 4.2 La Ferrassie (Les Eyzies, France). Left, engraved bone found with the adult skeleton in burial 1. Middle, the lower face, decorated with cup holes, of the stone slab that covered the burial pit of individual 6, a 3 to 5-year-old child. Right, plan and profile of the burial. *After Peyrony (1934).*

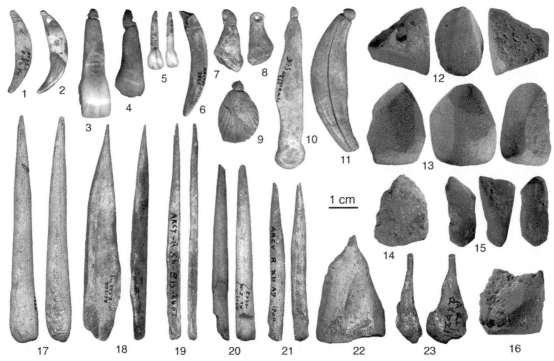

FIGURE 4.3 Grotte du Renne (Arcy-sur-Cure, France). Châtelperronian symbolic artefacts. Personal ornaments made of perforated and grooved teeth (1–6, 11), bones (7–8, 10) and a fossil (9); red (12–14) and black (15–16) colourants bearing facets produced by grinding; bone awls (17–23). *After Caron et al. (2011)*.

of time. Moreover, the reliability of the described stratigraphy has since been confirmed by limited excavation of a stratigraphic baulk carried out in 1998. These observations are relevant because the Châtelperronian levels yielded no less than 39 personal ornaments or fragments thereof (mostly pendants made of animal bone and teeth and of fossils), as well as ~18 kg of yellow, black and (mostly) red colourants (Fig. 4.3). At the time, these finds were significant in two ways. First, they represented the earliest evidence for symbolism anywhere, justifying Leroi-Gourhan's (1964) view of the Châtelperronian as representing the "dawn of art." Second, they were associated with human teeth of archaic affinities, suggesting that their creators could well have been the Neanderthals. This suspicion was eventually confirmed in 1979, with the discovery of a Neanderthal skeleton in a Châtelperronian context at the French site of St.-Césaire, in the Charentes (Lévêque and Vandermeeer, 1980), and later, for the Grotte du Renne itself, with the results of the comparative study of the human teeth and child temporal bone found in its Châtelperronian levels (Hublin et al., 1996; Bailey and Hublin, 2006).

Where complex technology is concerned, the key example comes from the open-air site of Königsaue. Located in the margins of the Aschlerlebener paleolake in Saxony-Anhalt, eastern Germany, this site was excavated in 1963 by Mania (2002). It yielded two fragments of birch bark pitch, one of which bore a human fingerprint as well as impressions of a flint blade and of wood-cell structures, indicating the use as an adhesive material to fix a wooden haft to a stone knife. Chemical analysis of these finds (directly dated by radiocarbon, in the meanwhile, to ≥50,000 years ago) eventually showed that the pitch had been produced through a several-hour-long smouldering process requiring a strict manufacture protocol – under exclusion of oxygen and at a tightly controlled temperature (between 340 and 400 °C) (Koller et al., 2001). Similar finds have since been made at the Italian site of Campitello, where they date to an even earlier period, >120,000 years ago (Mazza et al., 2006). This birch bark pitch is the first known artificial raw material in the history of humankind, and the fire technology underpinning its production is of a level of sophistication that was to remain unsurpassed until the invention of Neolithic pottery kilns.

Thus, by the late 1980s, when the Mitochondrial Eve hypothesis was put forth, there was sufficient evidence indicating that, under the criteria subsequently formalised by palaeoanthropologists working in Africa, the Neanderthals had been culturally modern, even if, biologically, as supporters of RAO would go on to argue, they had been a different species and had made no genetic contribution to subsequent generations of humans. It is important to note, in this context, that, at the time, nothing even remotely equivalent to the European Middle Palaeolithic finds of the

1930s–1960s existed in the archaeological record of the African Middle Stone Age (MSA). Until the end of the 20th century, in fact, the earliest secure evidence of symbolic material culture in the African continent was represented by the painted slabs of the Apollo 11 cave, dated to about 30,000 calendar years ago (Wendt, 1974, 1976).

Therefore, with the benefit of hindsight, it seems obvious that only the following alternatives represented logically consistent corollaries of the late 1980s knowledge base concerning Neanderthals and modern humans: either cultural evolution needed to be decoupled from biological evolution, with Neanderthals and Moderns representing morphological "species" of humans that, cognitively and behaviourally, were equally advanced, implying that such capabilities were already present in the common ancestor, or, if a direct link between cognition, behaviour and the underlying biological/genetic variation was postulated, then, on the strength of the empirical evidence, Neanderthals had to be considered the more advanced species.

Given this, the research questions that should have arisen as a by-product of Mitochondrial Eve would have to be as follows:

- in the first case, what explained the differential manifestation, in time and space, of biological capabilities with a very remote ancestry, and why did it take so long for them to become apparent in the archaeological record?
- in the second case, why were Neanderthals replaced by modern humans, if, on the strength of the archaeological evidence, they seemed to have been the behaviourally more advanced (maybe as a result of a biologically based cognitive superiority) "species"?

Instead, the questions that inspired research carried out under the human revolution umbrella for the subsequent quarter of a century were the following: Given that Neanderthals went extinct, early moderns must have been the superior "species," so, how do we explain away the European evidence? And, given that anatomical modernity emerged in Africa before 100,000 years ago, shouldn't we expect behavioural modernity to have emerged alongside and, if so, where is the evidence hiding, and how do we go about uncovering it?

4.3. NEANDERTHAL SYMBOLISM: THE LAST DECADE OF RESEARCH

It is undeniable that these human revolution questions were productive. They led to a spurt of research in Africa that eventually changed our knowledge of the continent's MSA rather dramatically, making possible the formulation of the definitions of behavioural modernity given above. The critical discoveries were perhaps (a) the *Nassarius kraussianus* shell beads and engraved ochre crayons from the Blombos Cave, in South Africa, (b) the decorated ostrich eggshell containers from the Diepkloof rockshelter, also in South Africa, (c) the *Nassarius gibbosulus* shell beads from the Grotte des Pigeons (Morocco) and a number of other sites in the Maghreb, (d) the heat pretreatment of the silcrete used for the manufacture of bifacially flaked points and blade/bladelet blanks documented at the Blombos Sands and Pinnacle Point sites of South Africa for the Still Bay and Howiesons Poort facies of the regional MSA and (e) the harpoon-like bone points from Katanda, Democratic Republic of the Congo, where such points are likely to have been used for the catching of large fish (Yellen, 1996, 1998; Henshilwood et al., 2004; Bouzouggar et al., 2007; Brown et al., 2009; Texier et al., 2010).

One could have expected these finds to have led the experts to conclude that the apparent conundrum of the late 1980s (Neanderthals, the behaviourally more advanced species, replaced by the less advanced Moderns) was simply an artefact of insufficient research. Simply put, as some of the evidence summarised in the preceding paragraph goes back to as much as 100,000 years ago, "out-of-Africa" and the "human revolution" could indeed work because symbolism in Africa was as early as, if not earlier than, symbolism in Europe. More often than not, however, this new evidence was portrayed as vindicating the notion that, from the beginning, Moderns had had what Neanderthals had never had, the demise of the latter thus becoming self-explanatory. As a result, the last decade of the 20th and the first decade of the 21st century were still characterised by intense, often acrimonious, debate concerning the reliability of the evidence for Neanderthal symbolism. For the most part, controversies concerned the interpretation of the Châtelperronian, and particularly the evidence from the Grotte du Renne, but included assessments of the archaeology of the Middle Palaeolithic of Europe that omitted other key evidence. For instance, where La Ferrassie is concerned, even in the framework of extensive, almost encyclopaedic reviews of the French Mousterian (e.g., Mellars, 1996), commentators generally ignored the existence of the decorated bone associated with individual 1, or dismissed the cup holes in the slab covering the grave pit of individual 6 as possibly but not certainly artificial. Some even went as far as denying the fact of burial itself, not only at La Ferrassie but among Neanderthals in general (Gargett, 1989, 1999).

Elsewhere (Zilhão, 2007, 2011), I have provided detailed discussions of the empirical and logical inconsistencies that I perceive in the various propositions put forth in the context of the human revolution to explain away the evidence for Neanderthal symbolism; so only a brief inventory is necessary here. Basically, such propositions fall into two families, ones that accept the reality of the archaeological association between symbolic material culture and Neanderthal-made stone tools and ones that reject it. Acculturation – the notion that the Neanderthals would have been led to adopt symbolic material culture, refashioned in their own terms, as a by-product of contact with immigrating moderns – was also proposed as an explanation for the Châtelperronian (Hublin, 2000), but this notion is not relevant here, as, by definition, it implies that the adopters possessed the same cognitive capabilities as those from whom the inspiration had been received.

The first family of explanations (e.g., Bordes, 1981; Stringer and Gamble, 1993; Hublin et al., 1996; Gravina et al., 2005; Mellars, 1999, 2005; Bar-Yosef, 2006; Bar-Yosef and Bordes, 2010) revolves around the notion that the association between symbolic material culture and Neanderthal-made stone tools is incidental and an artefact of long-term regional contemporaneity between late Neanderthal and early modern human cultures. For instance, the Châtelperronian could in fact have been made by modern humans, with the St.-Césaire skeleton representing the victim, not the maker, of the stone tools found in the deposit that contained it. Or fluctuating boundaries between the territories of the two groups might have led to situations where the remains of consecutive occupations of the same place by one and then the other became incorporated in an occupational palimpsest where the symbolic artefacts represented material left behind by modern humans, not Neanderthals. Alternatively, in the course of their perambulations across territory previously used by modern humans, Neanderthals could have come across abandoned material culture items that they collected as *curiosa* and brought back to their camps. Finally, the objects might be genuine Neanderthal craft but reflected imitation of behaviours observed among their modern human neighbours without any understanding of the deep symbolic meaning underlying the functional role they played in the societies of those neighbours. In a distant echo of King's (1864) comparative approach, the Neanderthals were even compared to children and to the "primitives" of 19th- and early 20th-century ethnography:

"... the replication of aeroplane forms in the New Guinea cargo cults hardly [implies] an understanding of aeronautics or international travel (...). To draw another analogy, if a child puts on a string of pearls, she is probably doing this to imitate her mother, not to symbolise her wealth, emphasise her social status, or attract the opposite sex" (Mellars, 1999; pp. 350).

The second family of explanations revolves around the contentions that no symbolism can, in fact, be inferred from the purported "symbolic" material culture documented among the Neanderthals or that the association of such material culture with Neanderthal-made stone tools is spurious and an artefact of postdepositional processes. Where burials are concerned (e.g., Gargett, 1989, 1999), the evidence was deemed equivocal and it was argued that purposeful protection is not necessary to account for the preservation of articulated skeletons, which could result from the operation of entirely natural processes. Where personal ornaments are concerned (e.g., White, 2001, 2002; Taborin, 2002), it was pointed out that most of the evidence comes from the Châtelperronian and even then practically from a single site, the Grotte du Renne, where the corresponding levels are overlain by Aurignacian ones – as the Aurignacian is modern human related, the simplest explanation for the anomaly would be that the personal ornaments found deeper in the sequence are intrusive items.

After much debate, the first family of explaining-away propositions can nowadays be safely put to rest (and, for the same empirical reasons, such is also the case with acculturation). The analysis of continent-wide chronostratigraphic patterns (Zilhão and d'Errico, 1999, 2003; Zilhão, 2006, 2007), coupled with improvements in radiocarbon dating that corrected erroneous results and corroborated the conclusions derived from chronostratigraphy (e.g., Higham et al., 2009), has shown that the emergence of the Châtelperronian occurred sometime between 45,000 and 43,000 years ago. Therefore, it predates by many millennia both the Aurignacian (the earliest dates for which are of some 41,500 years ago) and the earliest unambiguous skeletal evidence for anatomical modernity anywhere in the continent (which, at present, is represented by the two Oase fossils from Romania, the Oase 1 mandible, directly dated to about 40,000 years ago, and the Oase 2 cranium; Trinkaus et al., 2003; Rougier et al., 2007). Therefore, even if one were to admit that the specific associations seen at St.-Césaire and the Grotte du Renne are open to question, the corollary that the Châtelperronian could have been made by modern humans instead of Neanderthals would be valid only if proof were to be provided that modern humans were present in Europe well before the time of the Oase fossils.

Recently, claims that such was the case have been forthcoming indeed (e.g., Benazzi et al., 2011; Higham et al., 2011), but so far they remain unsupported. Higham et al.'s argument is that the Kent's Cavern maxilla, in whose teeth they recognise modern human affinities, dates to the 43rd millennium. However, they did not date the fossil itself. The age estimate is based on the presumed stratigraphic association of the maxilla with some faunal remains that they did date, but the finds come from a context that was very poorly excavated and where recent archaeological work identified severe stratigraphic disturbance (Pettit & White, 2012); in fact, the specimen could well be of much more recent age, as eventually was shown to be the case with a significant number of human remains once thought to date to the time of the Middle-to-Upper Palaeolithic transition (most famously those from the German cave site of Vogelherd; Conard et al., 2004). Benazzi et al., in turn, argue that two deciduous molars from level E-III of Grotta del Cavallo (Nardò, Italy), placed beyond 43,000 years ago by dates for the overlying strata, are of modern humans. However, a number of studies have shown that, in tissue organisation as much as in external morphology, the overlap between Neanderthals and modern humans is such that secure assignment of isolated teeth to one or the other is simply not possible. For instance, Bayle et al. (2010) have shown that, in the dentition of a single individual, some teeth can present the "Neanderthal" endostructural pattern and other teeth the "modern human" pattern, while the statistical method developed by Bailey et al. (2009) on the basis of a sample of 158 teeth securely associated with Neanderthals and Upper Palaeolithic modern humans classified the Oase 1 mandibular dentition as modern and the Oase 2 maxillary dentition as Neanderthal.

Where the second family of explaining-away propositions is concerned, results from the recent re-analysis of the context of the child skeleton excavated in 1961 at Roc-de-Marsal (Dordogne, France) would seem, at first glance, to provide supporting empirical evidence (Sandgathe et al., 2011). However, even if this particular instance of a Neanderthal burial were to be rejected it does not necessarily exclude all other contenders. The key argument here is one that Leroi-Gourhan (1964) had already advanced more than two decades before the onset of the 1990s debate on whether Neanderthals had indeed buried their dead: If the large number of articulated human skeletons featured by the archaeological record of the Middle Palaeolithic is not a reflection of the emergence of intentional burial, how then do we explain (a) why identical instances of articulated human skeletons remain unknown from earlier levels of similar (if not the same) sites and (b) why do we not find identical instances of articulated skeletons of other cave-dwelling animals (foxes, wolves, hyenas, etc.) in the same deposits? In any case, intentional interment is documented by the ongoing re-excavation work carried out at the La Chapelle-aux-Saints complex of cave and rockshelter sites. This work has corroborated the original excavators' accounts and was even able to re-expose the actual burial pit described by them at the Bouffia Bonneval, the 1908 site where the old man's skeleton was found, thereby confirming the ontological reality of both the feature and the behaviour that produced it (Rendu et al., 2011).

The notion that the personal ornaments from the Châtelperronian of the Grotte du Renne are intrusive is also completely inconsistent with their vertical distribution across the site's stratigraphic sequence: two-thirds of the combined total for the site's Aurignacian and Châtelperronian levels came from the lowermost Châtelperronian, and only 17% were found in the levels whence all the others had been putatively displaced. No less an inconsistency is apparent when the notion is contrasted with the vertical distribution of the index fossils of the relevant technocomplexes: not one out of 287 Dufour bladelets and not one of their 2800 unretouched blanks (Aurignacian diagnostics recovered in level VII) moved down into the Châtelperronian (found in levels VIII–X), while only one out of 385 Châtelperron points (diagnostic of the Châtelperronian) moved up into the Aurignacian. This pattern confirms the overall stratigraphic integrity of the site, as one could easily surmise from the good preservation of habitation features in the basal Châtelperronian deposit, where the spatial distribution of ornaments, worked bone, pigments and pigment-processing tools is congruent with the location of those features (Caron et al., 2011; Fig. 4.4).

Moreover, the Grotte du Renne is not alone. In France, other Châtelperronian sites have yielded similar types of ornaments, namely the rockshelter of Quinçay, in the Charentes (Fig. 4.5), where intrusion from overlying Aurignacian levels is hard to defend, as such levels are nonexistent at the site and the Châtelperronian deposits are sealed by collapsed limestone slabs several metres long and tens of centimetres thick (Roussel and Soressi, 2010). In Germany, the find horizon Ranis 2 of the Ilsenhöhle, a collapsed rockshelter excavated in the 1930s near Ranis, Thuringia (Hülle, 1977), yielded an ivory disc with a central hole, as well as a needle-like bone point. The associated stone tools have Altmühlian/Szeletian affinities, i.e., belong in a technocomplex characterised by the production of fine bifacial foliates that is found across southern Germany, Moravia and southern Poland. Across its area of occurrence, this particular type of stone tool production is radiocarbon- or stratigraphically-dated to before the Aurignacian and to about the same time interval as the Châtelperronian. In Belgium, 19th-century excavators working at the site of Trou Magrite (Pont-à-Lesse) found an ivory ring identical to those from the Châtelperronian of the Grotte du Renne. The associated lithics form a mixed collection where three different components (late Mousterian, Altmühlian/Szeletian and Aurignacian) can be recognised and, against previously held views (Otte, 1979; Lejeune, 1987; Moreau, 2003), the regional setting now favours the hypothesis that this object relates not to the Aurignacian but to one of the other, Neanderthal-associated occupations of the site. In Bulgaria, three items were recovered in level 11 of the Bacho Kiro cave (Dryanovo), the type site of the Bachokirian, which is broadly contemporary with the Châtelperronian: a spindle-shaped bone pendant that is oval in cross section and grooved at the narrow end, and fragments of two pierced teeth from unidentified species (Kozłowski, 1982). Finally, in central and eastern Mediterranean Europe, the Uluzzian technocomplex, now firmly dated to the time range of the Châtelperronian at the cave site of Klisoura 1 (Prosymna, Greece) and at the rockshelter of Fumane (Molina, Italy), features large numbers of shell beads, mostly *Dentalium* sp.

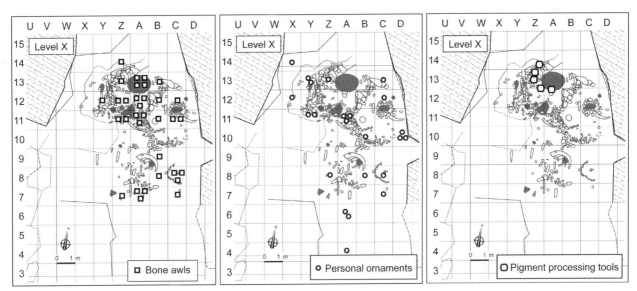

FIGURE 4.4 Grotte du Renne (Arcy-sur-Cure, France). Distribution of bone awls, ornaments and pigment-processing tools in Châtelperronian level X. The areas in grey are hearths. *After Caron et al. (2011).*

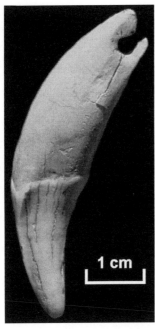

FIGURE 4.5 La Grande Roche de la Plématrie (Quinçay, France). Perforated wolf canine from the Châtelperronian. *After Zilhão and d'Errico (1999).*

tubes (Gioia, 1990; Palma di Cesnola, 1993; Koumouzelis et al., 2001a, 2001b; Higham et al., 2009; Peresani et al., 2011).

In each case, these technocomplexes represent the initial Upper Palaeolithic of the corresponding regions, ranging in calendric age between ~41,000 and ~45,000 years ago. Recently, however, evidence has been produced that personal ornamentation was in existence in the Neanderthal world even during the preceding Middle Palaeolithic. At Cueva de los Aviones (Cartagena, Spain) and Cueva Antón (Mula, Spain), four types of finds were made in Mousterian levels dating to as early as ~50,000 years ago (Zilhão et al., 2010a; Figs. 4.6 and 4.7): perforated shells of large marine bivalves of the genera *Acanthocardia*, *Glycymeris* and *Pecten*, some of which are painted; one unperforated upper valve of the Mediterranean spiny oyster, *Spondylus gaederopus* (characterised by exuberant sculpture and vivid red or violet colour, two features that inspired collection for ritual purposes in a large number of archaeological and ethnographic contexts worldwide), which had been used as a container for the storage or preparation of a complex cosmetic recipe whereby shiny bits of freshly ground hematite and pyrite (black) were added to a base of lepidocrocite (red); lumps of iron pigments of different mineral species (hematite, goethite and siderite), but mostly of yellow natrojarosite (whose only known use is in cosmetics); and a kind of stiletto made of an unmodified pointed bone bearing pigment residues on the broken tip, suggesting use in the preparation or application of colourants.

There can be little doubt that the straightforward interpretation of this Spanish material is that the pigments were used in bodily, most likely facial, decoration, and the perforated shells in personal ornamentation, probably as neck pendants. Body painting has also been inferred for the crayons of black manganese found at the Mousterian site of Pech de l'Azé (Carsac-Aillac, France), and in this case such an interpretation is supported by experimental replication and use-wear analysis (Soressi and d'Errico, 2007). Like those from Cueva de los Aviones, the similarly perforated and ochred *Glycymeris* sp. shells from the Middle

FIGURE 4.6 Cueva de los Aviones (Cartagena, Spain). Perforated shells from Mousterian level II: 1. *Acanthocardia tuberculata*; 2-3. *Glycymeris insubrica*. Remains of red pigment (hematite) were found adhering to the inner side of no. 3. *After Zilhão et al. (2010a)*.

Palaeolithic of Qafzeh, in Israel (Bar-Yosef et al., 2009), have also been interpreted as personal ornaments (in this case, in a modern human-related context and, as one would expect, rather uncontroversially). This evidence from the Eurasian Middle Palaeolithic has since been further strengthened by the finding in the Mousterian levels of Fumane of cut-marked bones of large birds of prey whose skeletal provenance, processing patterns and taphonomic context leave little doubt that they stand for the intentional extraction of feathers for ornamental purposes (Peresani et al., 2011).

In short, the makers of the Châtelperronian and coeval European technocomplexes must have been the Neanderthals simply because there was nobody else around in Europe at that time; the presence of body painting and personal ornamentation in those technocomplexes' cultural repertoire is demonstrated by the association of the corresponding artefacts with diagnostic stone tools in archaeological contexts whose integrity has passed the test of intensive scrutiny; similar, earlier evidence is now known from the Middle Palaeolithic; so no reason exists to treat such associations as problematic to begin with.

4.4. ONGOING CONTROVERSIES: WHY?

Overall, these recent developments have met widespread acceptance, among both academics and the general public, which is probably at least in part related to the fact that the first results of the Neanderthal genome project, published at about the same time (Green et al., 2010), corroborated the palaeontology- and archaeology-based assimilation model of Neanderthal "extinction" (e.g., Smith et al., 2005; Trinkaus, 2007). The realization that the time of contact witnessed significant interbreeding between aboriginal Neanderthals and immigrating modern humans removed the rationale for thinking about that time in terms of different, competing species, rendering fully human cognition the null hypothesis for how Neanderthal brains worked and making expressions of fully symbolic material culture in their archaeological record the thing to be expected rather than an anomaly to be explained away.

Or so it should have been. However, while many former and prominent supporters of the human revolution have indeed taken the evidence on board and moved on to ask new questions arising out of this scientific watershed (e.g., Watts, 2010; d'Errico and Stringer, 2011), the old ways not only did not die out but continue to undergo (sometimes rather potent) bursts of expression. The invited PNAS commentary on recently obtained radiocarbon results for the Grotte du Renne, for instance, stated that their "central and inescapable implication" was that "the single most impressive and hitherto widely cited pillar of evidence for the presence of complex 'symbolic' behaviour among the late Neanderthal populations in Europe has now effectively collapsed" (Mellars, 2010; pp. 20148). The basis for this extraordinary claim resided in the fact that the dating results for the site's Châtelperronian levels (Higham et al., 2010) ranged from ~21,000 to ~49,000 radiocarbon years ago. This wide age range was taken to imply two things: (a) a substantial degree of stratigraphic mixing and (b) derivation from the immediately overlying Aurignacian of a significant proportion of the dated samples and, by inference, of the personal ornaments found in the Châtelperronian levels. Since the authors of the dating study had themselves cautiously flirted with such implications, the commentary was not entirely out of place, but, from an empirical point of view, were those implications in any way justified?

FIGURE 4.7 Two versions of the same concept, the ornamental use of perforated *Pecten* shells, both abandoned as half-valves after breakage, in the Middle and Upper Palaeolithic of the Mula basin (Murcia, Spain). Top: *P. jacobaeus* from Middle Gravettian level 7 of the rockshelter of Finca Doña Martina (Zilhão et al. 2010b). Bottom: *P. maximus* from Mousterian level I-k of Cueva Antón (Zilhão et al., 2010a). In both shells, the external side is painted, more apparently so in the Cueva Antón specimen, where the natural red color had been lost due to bleaching; pigment is also visible in the internal side of the Gravettian shell. The distance between the two sites is 2.5 km and they feature functionally similar kinds of human occupations. The symbolic significance of such shells, collected from the seaside, >60 km away, is uncontroversial when found in modern human-associated Upper Palaeolithic contexts, and there is no reason to think otherwise when they are found in the Neandertal-associated Middle Palaeolithic.

The first thing to bear in mind when assessing the Grotte du Renne results is that, technically, dating the site's Châtelperronian has been an extremely challenging exercise. Given regional culture-stratigraphic patterns, its age ought to be in excess of 36,500 radiocarbon years (i.e., 41,500 calendar years), as repeatedly pointed out for almost a decade now (e.g., Zilhão and d'Errico, 2003). Prior to this last attempt at dating it, however, only 2 out of the 17 (12%) available results satisfied that condition. For a long time, the other results were the single pillar of evidence supporting the "acculturation" and "imitation without understanding" explanations of Châtelperronian personal ornaments. However, they were patently anomalous when viewed in their proper context, and there were repeated warnings that they were likely to be no more than minimum ages and an artefact of poor collagen preservation at the site (Zilhão, 2006, 2011).

In this regard, a dating experiment carried out at the Sesselfelsgrotte, in Bavaria, Germany (Richter, 2002), had already provided a pertinent cautionary tale. Here, bone samples collected from exposed areas of the site (exterior or close to the drip line, where the deposits had undergone long-term postdepositional leaching) systematically yielded much younger ages than those from interior, well-protected areas of the same levels. Knowing that overhang collapse soon after the Châtelperronian occupation transformed the corresponding levels of the Grotte du Renne into essentially an open site for more than thirty millennia, the example from Sesselfelsgrotte makes it clear that collagen degradation and incomplete decontamination producing erroneously young results are the parsimonious explanation for the dating anomalies seen at the French site. Another German example of a situation akin to the Grotte du Renne is the Ilsenhöhle, where all four dates for the Ranis 2 Altmühlian/Szeletian find horizon came out too young, one of them even in the Magdalenian range (Grünberg, 2006). Although a Magdalenian occupation of the site (the Ranis 4 find horizon) exists in levels >2 m higher up in the sequence, the cultural coherence of the Ranis 2 stone tool assemblage indicates that incomplete decontamination, not postdepositional disturbance, is likewise the reasonable explanation for the dating anomalies.

The new results represent a significant improvement of the Grotte du Renne situation, as 13 out the 21 Châtelperronian samples (62%) obtained by Higham et al. (2010) satisfy the condition that they should yield ages in excess of 36,500 radiocarbon years before present. This improvement is related to major developments in pretreatment and measurement, namely, the use of ultra-filtration (Higham, 2011). However, given previous history and the fact that poor preservation was confirmed by the failure of 19 out of 50 samples used in the experiment, can we be confident that the improved techniques managed to completely solve the site's contamination problems? In my view, we cannot. Therefore, under Occam's razor, the past dating history of the site and the pattern of overall stratigraphic integrity apparent in the vertical and horizontal distributions of key finds and features indicate that the presence of residual contaminants is a much better explanation for the anomalously young results yielded by some samples than their reflecting significant post-depositional disturbance of the sequence. In fact, no natural mechanism could possibly move 38% of the samples down from the Aurignacian to the Châtelperronian while leaving *in situ* in the Aurignacian level all of its bladelets (Caron et al., 2011).

When proper attention is paid to the details of the dating study, the suspicion that the problem lies in the dating rather than in the stratigraphy is considerably strengthened. For instance, 84% of the results obtained came from samples that either had been treated with glues and consolidants or were suspected of having been so treated – hardly the ideal kind of sample upon which to decide the major issues raised by the Grotte du Renne's record. Also, two of those results on consolidated samples did feature C:N (carbon:nitrogen) ratios above the laboratory's normal threshold for acceptance (3.5), indicating the presence of exogenous carbon, i.e., contamination. Finally, if we consider only the five samples for which treatment with consolidants was neither documented nor suspected, the results do come out in perfect stratigraphic order.

In short, it would seem that the news of the bursting of the "Neanderthal symbolism and ornament manufacture bubble" (Mellars, 2010) was, as Mark Twain would have put it, greatly exaggerated. The Oxford laboratory developed or fine-tuned many of the technical innovations that have allowed the radiocarbon dating of the transition to move forward so significantly, but why did this laboratory's researchers choose to validate samples and results that, in any other context, would have been either rejected outright or at least treated with great suspicion? Also, even within their interpretive framework, why did they choose to emphasise that the glass was one-third empty instead of emphasising that it was two-thirds full? I can think of no explanation other than the enduring influence in academia of perceptions of Neanderthals inherited from the Victorian age that predispose many scholars to readily accept anything that goes along with such perceptions, and to resist anything that goes against them with levels of scepticism that go way beyond those required by the scientific method.

Although scientists like to think of themselves as unprejudiced and independent-minded, a recent anecdote illustrates well how even the most respected sancta sanctorum of academia are affected by the pervasive influence of their cultural environment. In the spring of 2009, the publication of an ivory statuette from the German Aurignacian (Conard, 2009) made headlines the world over. The buzz was justified, as the object was the oldest female figurine known so far, and one of the oldest examples (if not the oldest) of figurative art with good contextual and dating evidence. The exuberant bust and other well-marked features of the female body also made it entirely predictable that the tabloid press would promote it the way it did – in the words of *The Sun* (issue of May 14, 2009), as the world's first ever "Page 3 girl" (http://www.thesun.co.uk/sol/homepage/news/2427906/A-CURVY-statuette-35000-years-old-has-been-unveiled-by-scientists-as-the-worlds-earliest-model-of-the-female-bodyThe-ivory-figurine-with-big-boobs-could-be-the-first-ever-Page-3-girl.html, accessed May 9, 2011). Perhaps less predictable, however, was that much the same line was followed by *Science Now* (issue of May 13, 2009), the news portal of the American Association for the Advancement of Science, which commented on the find in a piece under the title "The Earliest Pornography?" (http://news.sciencemag.org/sciencenow/2009/05/13-01.html, accessed May 9, 2011). Concurring statements from reputed scholars were quoted, as well as the "nothing's changed in 40,000 years" reaction of one of those scholars' male colleague to whom the figurine had been shown.

Perhaps this 21st century male was right, but is it not at least equally plausible that the figurine had in fact been made and worn by a woman, and had nothing to do with the "sex-madness" attributed to the average Aurignacian male by another of those commentators? Could it be that this was simply another instance of the drive to see the Aurignacians as proper moderns "just like us," as opposed to the improper, "quite not like us" Neanderthals that preceded them in the same regions? Also, since we are at it, could it be that attitudes towards the Châtelperronian personal ornaments of the Grotte du Renne are likewise influenced by this attitude more than by the dispassionate, purely "natural science" assessment of the site's stratigraphy and dating? Since, over the last quarter of a century, this is what happened time and again whenever the symbolism of Neanderthal material culture seemed to be the obvious implication of the then-available evidence, chances are that such is indeed the case in this instance.

The scattered nature of the evidence concerning body ornaments, the fact that it comes from a very small number of sites and questions of association with the human remains found in the same levels at the key locality (the Grotte du Renne) are often raised as objections to the acceptance of fully symbolic behaviour among Europe's later Neanderthal populations. In a glaring example of the double standards issue, no such "rarity" objections, however, have been raised in relation to the evidence from the African MSA, where, in all these regards, there is much stronger ground for such objections. Blombos remains the single South African site to have yielded perforated shell beads securely associated with well-dated MSA occupations, no human remains were found alongside and the closest in time are the slightly earlier fossils from Klasies River Mouth, whose morphology is archaic rather than modern (Trinkaus, 2005). Where the Maghreb is concerned, the so-called Dar-es-Soltan people living in the area at the time of production of the perforated *Nassarius* shells found at a number of Aterian sites are, morphologically, archaic, not modern (Klein, 1992; Trinkaus, 2005). In Eastern Africa, where, according to RAO, anatomically modern humans first sprung into being and whence they spread into the rest of Africa, known MSA sites have so far failed to yield similar evidence. None of this has prevented most palaeoanthropologists from accepting that Blombos and the Maghreb sites represent the "modern behavior" of "modern humans" in the African continent as a whole. Yet, eyebrows are raised concerning Europe, where the "modern behavior" of "archaic humans" is supported by the immediate association in the same level, of the same site, between diagnostic Neanderthal fossils, large numbers of personal ornaments and amounts of colourants considerably exceeding those so far reported for any coeval African site!

4.5. CONCLUSION

One thing is certain after 150 years of Neanderthal debates: the answer to the questions above will remain open for quite sometime. To my mind, however, present evidence dictates that, like evolution, Neanderthal symbolism should be treated as fact, not hypothesis. Given that in and of itself Neanderthal-ness implies hundreds of thousands of years of divergent evolutionary trajectories between Africa and Europe – even if gene flow was never interrupted and, as pointed out by Holliday (2006), the isolation was insufficient in extent and duration to result in speciation – the corollary of this conclusion is that the emergence of symbolism cannot relate to processes of genetic or other biologically based, "flick-of-a-switch" change occurring in a geographically restricted, small but subsequently expanded, population. In this context, the only realistic models are those that explain symbolic material culture as a by-product of the increasing complexity of social interactions, resulting from demographic growth facilitated by technological progress and increasing adaptive success, and reaching a threshold at about the same time interval across vast regions of the Old World (e.g., Gilman, 1984; Shennan, 2001; Powell et al., 2009). Only such a process can explain both the geographical unevenness of the emergence of material symbols and the "now you see it, now you don't" pattern (Hovers and Belfer-Cohen, 2006) displayed – as befits the beginning of any curve of exponential growth – by its initial stages, in Europe as much as in Africa.

In this context, an interesting possibility is raised by the evidence for symbolic material culture among Neanderthals as far back as the Middle Palaeolithic, ≥50,000 years ago (as documented at La Ferrassie, Cueva de los Aviones and Fumane). In the Near East of last interglacial times, the presence of the African tradition of *Nassarius* beads so far rests on a single find from the cave of Skhul (Israel). Although Vanhaeren et al. (2006) made a good case for that bead to come from the level that contained the burials of early modern humans, the true age of the burials themselves is controversial. Uranium-series dates obtained on animal teeth suggest that an important component of that deposit dates to only 40,000–45,000 years ago (McDermott et al., 1993; Grün et al., 2005), and the palaeoanthropologists of different persuasions who have studied the skeletons (e.g., Stringer, 1998) seem to concur that two chronologically distinct populations, one anatomically less "modern" than the other, could well be represented in the Skhul sample. The possibility exists, therefore, that the Skhul *Nassarius* bead relates to a later period of occupation, i.e., to the modern humans who returned to the Near East after 45,000 years ago, not to those who lived there during the last interglacial, whose only items of personal ornamentation would therefore be the perforated-ochred *Glycymeris* sp. shells associated with the burials of that age found at the nearby cave site of Qafzeh.

Bearing in mind the long duration of traditions of body decoration (Stiner, 1999), the fact that this significant cultural trait – the ornamental use of perforated/painted, bivalve shells – is shared between the Qafzeh Moderns and the Aviones Neanderthals suggests the following hypothesis: that the Aviones shells represent the persistence in Europe, among Neanderthal societies, of traditions of personal ornamentation going back to the last interglacial, at which time they would have been spread around at least the eastern and northern Mediterranean seaboards, regardless of (real or perceived) biological boundaries. Put another way, it is possible that, some 90,000 years ago, two different ornament traditions were already in existence: one in Africa associated with modern humans or their immediate ancestors – the *Nassarius* beads tradition of the Still Bay culture of South Africa and the Aterian culture of the Maghreb; another in Mediterranean Europe and the Near East associated with both modern humans and Neanderthals – the tradition of pendants made of nonfood, large bivalve shells documented by the Tabun C-type Mousterian of Qafzeh and the Middle Palaeolithic of Iberia. In this context, the Châtelperronian and its personal ornaments made of animal bones and teeth pierced or grooved for suspension would represent yet a third tradition, one whose origins may well be more recent and possibly related to social and demographic processes triggered by the northward expansion of humans in Europe after the end, ~60,000 years ago, of the first cold phase of the last glaciation.

Future research will show whether these hypotheses will be supported by the evidence, and whether such hypothesised ornamental traditions relate, even if in an incomplete and perhaps distorted way, to the cultural markers of geographic significance and time depth that Richter (2000) sees in the typology and technology of the stone tools of the Middle Palaeolithic of the Greater Mediterranean region. Here, they serve to illustrate the kinds of questions that should lie ahead of us, now that "the Neanderthal problem", to take up the title of a late 20th century *Current Anthropology* discussion forum (Fox, 1998), can be recognised for what it really is: a textbook example of the popular saying that "you never have a second chance to make a first impression."

REFERENCES

Bailey, S.E., Hublin, J.J., 2006. Dental remains from the Grotte du Renne at Arcy-sur-Cure (Yonne). J. Hum. Evol. 50, 485–508.

Bailey, S.E., Weaver, T.D., Hublin, J.J., 2009. Who made the Aurignacian and other early upper Paleolithic industries? J. Hum. Evol. 57, 11–26.

Bar-Yosef, O., 2006. Neanderthals and modern humans: a different interpretation. In: Conard, N.J. (Ed.), When Neanderthals and Modern Humans Met. Kerns Verlag, Tübingen, pp. 467–482.

Bar-Yosef, O., Bordes, J.-G., 2010. Who were the makers of the Châtelperronian culture? J. Hum. Evol. 59, 586–593.

Bayle, P., Macchiarelli, R., Trinkaus, E., Duarte, C., Mazurier, A., Zilhão, J., 2010. Dental maturational sequence and dental tissue proportions in the early Upper Paleolithic child from Abrigo do Lagar Velho, Portugal. Proc. Natl. Acad. Sci. U.S.A. 107, 1338–1342.

Benazzi, S., Douka, K., Fornai, C., Bauer, C.C., Kullmer, O., Svoboda, J., Pap, I., Mallegni, F., Bayle, P., Coquerelle, M., Condemi, S., Ronchitelli, A., Harvati, K., Weber, G.W., 2011. Early dispersal of modern humans in Europe and implications for Neanderthal behaviour. Nature. http://dx.doi.org/10.1038/nature10617.

Bordes, F., 1968. Le Paléolithique Dans Le Monde. Hachette, Paris.

Bordes, F., 1981. Un néandertalien encombrant. La Recherche 12, 643–645.

Boule, M., 1913. L'homme Fossile De La Chapelle-aux-Saints. Masson, Paris.

Bouzouggar, A., Barton, N., Vanhaeren, M., d'Errico, F., Collcutt, S., Higham, T., Hodge, E., Parfitt, S., Rhodes, E., Schwenninger, J.-L., Stringer, C., Turner, E., Ward, S., Moutmir, A., Stambouli, A., 2007. 82,000-year-old shell beads from North Africa and implications for the origins of modern human behavior. Proc. Natl. Acad. Sci. U S A 104, 9964–9969.

Brown, K.S., Marean, C.W., Herries, A.I.R., Jacobs, R., Tribolo, C., Braun, D., Roberts, D.L., Meyer, M.C., Bernatchez, J., 2009. Fire as an engineering tool of early modern humans. Science 325, 859–862.

Caron, F., d'Errico, F., Del Moral, P., Santos, F., Zilhão, J., 2011. The reality of Neandertal symbolic behavior at the Grotte du Renne, arcy-sur-cure. PLoS ONE 6 (6), e21545. http://dx.doi.org/10.1371/journal.pone.0021545.

Conard, N.J., 2009. A female figurine from the basal Aurignacian of Hohle Fels Cave in southwestern Germany. Nature 459, 248–252.

Conard, N.J., Grootes, P.M., Smith, F.H., 2004. Unexpectedly recent dates for human remains from Vogelherd. Nature 430, 198–201.

d'Errico, F., Henshilwood, Ch., Lawson, G., Vanhaeren, M., Tillier, A.-M., Soressi, M., Bresson, F., Maureille, B., Nowell, A., Lakarra, J., Backwell, L., Julien, M., 2003. Archaeological evidence for the emergence of language, symbolism, and music — an alternative multidisciplinary perspective. J. World Prehist 17, 1–70.

d'Errico, F., Stringer, C.B., 2011. Evolution, revolution or saltation scenario for the emergence of modern cultures? Philos. T R Soc. B 366, 1060–1069.

d'Errico, F., Zilhão, J., Baffier, D., Julien, M., Pelegrin, J., 1998. Neanderthal acculturation in Western Europe? a critical review of the evidence and its interpretation. Curr. Anthropol. 39 (Supplement), S1–S44.

David, F., Connet, N., Girard, M., Lhomme, V., Miskovsky, J.-C., Roblin-Jouve, A., 2001. Le Châtelperronien de la grotte du Renne àArcy-sur-Cure (Yonne). Données sédimentologiques et chronostratigraphiques. Bulletin de la Société Préhistorique Française 98, 207–230.

Davidson, I., Noble, W., 1996. Human Evolution, Language and Mind. A Psychological and Archaeological Inquiry. Cambridge University Press, Cambridge.

Defleur, A., 1993. Les Sépultures Moustériennes. CNRS, Paris.

Fox, R.G., 1998. Agonistic science and the Neanderthal problem. Curr. Anthropol. 39 (Supplement), S1.

Gargett, R., 1989. The evidence for Neandertal burial. Curr. Anthropol. 30, 339–341.

Gargett, R., 1999. Middle Palaeolithic burial is not a dead issue: the view from Qafzeh, Saint-Césaire, Kebara, Amud, and Dederiyeh. J. Hum. Evol. 37, 27–90.

Gilman, A., 1984. Explaining the upper Palaeolithic revolution. In: Spriggs, M. (Ed.), Marxist Perspectives in Archaeology. Cambridge University Press, Cambridge, pp. 115–126.

Gioia, P., 1990. An aspect of the transition between middle and upper Palaeolithic in Italy: the Uluzzian. In: Farizy, C. (Ed.), Paléolithique Moyen Récent et Paléolithique Supérieur Ancien en Europe, pp. 241–250. Mémoires du Musée de Préhistoire de l'Ile de France 3, Nemours.

Gravina, B., Mellars, P., Bronk Ramsey, C., 2005. Radiocarbon dating of interstratified Neanderthal and early modern human occupations at the Chatelperronian type-site. Nature 438, 51–56.

Green, R.E., Krause, J., Briggs, A.W., Maricic, T., Stenzel, U., Kircher, M., Patterson, N., Li, H., Zhai, W., Fritz, M.H.-Y., Hansen, N.F., Durand, E.Y., Malaspinas, A.-S., Jensen, J., Marques-Bonet, T., Alkan, C., Prüfer, K., Meyer, M., Burbano, H.A., Good, J.M., Schultz, R., Aximu-Petri, A., Butthof, A., Höber, B., Höffner, B., Siegemund, M., Weihmann, A., Nusbaum, C., Lander, E.S., Russ, C., Novod, N., Affourtit, J., Egholm, M., Verna, C., Rudan, P., Brajkovic, D., Kucan, Ž., Gušic, I., Doronichev, V.B., Golovanova, L.V., Lalueza-Fox, C., de la Rasilla, M., Fortea, J., Rosas, A., Schmitz, R.W., Johnson, L.F., Eichler, E.E., Falush, D., Birney, E., Mullikin, J.C., Slatkin, M., Nielsen, R., Kelso, J., Lachmann, M., Reich, D., Pääbo, S., 2010. A draft sequence of the Neandertal genome. Science 328, 710–722.

Grün, R., Stringer, C., McDermott, F., Nathan, R., Porat, N., Robertson, S., Taylor, L., Mortimer, G., Eggins, S., McCulloch, M., 2005. U-series and ESR analyses of bones and teeth relating to the human burials from Skhul. J. Hum. Evol. 49, 316–334.

Henshilwood, C., d'Errico, F., Vanhaeren, M., Van Niekerk, K., Jacobs, Z., 2004. Middle Stone Age shell beads from South Africa. Science 304, 404.

Henshilwood, C., Marean, C., 2003. The origin of modern human behavior. Critique of the models and their test implications. Curr. Anthropol. 44, 627–651.

Higham, T., Compton, T., Stringer, C., Jacobi, R., Shapiro, B., Trinkaus, E., Chandler, B., Gröning, F., Collins, C., Hillson, S., O'Higgins, P., Fitzgerald, C., Fagan, B., 2011. The earliest evidence for anatomically modern humans in northwestern Europe. Nature. http://dx.doi.org/10.1038/nature 10484.

Higham, T.F.G., 2011. European Middle and Upper Palaeolithic radiocarbon dates are often older than they look: problems with previous dates and some remedies. Antiquity 85, 235–249.

Higham, T.F.G., Brock, F., Peresani, M., Broglio, A., Wood, R., Douka, K., 2009. Problems with radiocarbon dating the Middle to Upper Palaeolithic transition in Italy. Quaternary Sci. Rev. 28, 1257–1267.

Higham, T.F.G., Jacobi, R.M., Julien, M., David, F., Basell, L., Wood, R., Davies, W., Bronk Ramsey, C., 2010. Chronology of the Grotte du

Renne (France) and implications for the context of ornaments and human remains within the Châtelperronian. Proc. Natl. Acad. Sci. U S A 107, 20234–20239.

Holliday, T.W., 2006. Neanderthals and modern humans: an example of a mammalian syngameon? In: Harvati, K., Harrison, T. (Eds.), Neanderthals Revisited: New Approaches and Perspectives. Springer, New York, pp. 289–306.

Hovers, E., Belfer-Cohen, A., 2006. "Now you see it, now you don't"—modern human behavior in the middle paleolithic. In: Hovers, E., Kuhn, S.L. (Eds.), Transitions before the Transition. Evolution and Stability in the Middle Paleolithic and Middle Stone Age. Springer, New York, pp. 295–304.

Hublin, J.-J., 2000. Modern-nonmodern hominid interactions: a Mediterranean perspective. In: Bar-Yosef, O., Pilbeam, D. (Eds.), The Geography of Neandertals and Modern Humans in Europe and the Greater Mediterranean. Peabody Museum Bulletin 8, Cambridge MA, pp. 157–182.

Hublin, J.-J., Spoor, F., Braun, M., Zonneveld, F., Condemi, S., 1996. A late Neanderthal associated with upper Palaeolithic artefacts. Nature 381, 224–226.

Hülle, W.M., 1977. Die Ilsenhöhle unter Burg Ranis/Thüringen. Eine Paläolitische Jägerstation. Gustav Fischer, Stuttgart.

King, W., 1864. The reputed fossil man of the Neanderthal. Q. J. Sci. 1, 88–97.

Klein, R.G., 2003. Whither the Neanderthals? Science 299, 1525–1527.

Klein, R.G., 1992. The archaeology of modern human origins. Evol. Anthropol. 1, 5–14.

Koller, J., Baumer, U., Mania, D., 2001. High-tech in the middle Palaeolithic: Neandertal-manufactured pitch identified. Eur. J. Archaeol. 4, 385–397.

Koumouzelis, M., Ginter, B., Kozłowski, J.K., Pawlikowski, M., Bar-Yosef, O., Albert, R.M., Litynska-Zajac, M., Stworzewicz, E., Wojtal, P., Lipecki, G., Tomek, T., Bochenski, Z.M., Pazdur, A., 2001a. The early upper Palaeolithic in Greece: the excavations in Klisoura cave. J. Archaeol. Sci. 28, 515–539.

Koumouzelis, M., Kozłowski, J.K., Escutenaire, C., Sitlivy, V., Sobczyk, K., Valladas, H., Tisnerat-Laborde, N., Wojtal, P., Ginter, B., 2001b. La fin du Paléolithique moyen et le début du Paléolithique supérieur en Grèce: la séquence de la Grotte 1 de Klissoura. L'anthropologie 105, 469–504.

Kozłowski, J. (Ed.), 1982. Excavation in the Bacho Kiro Cave (Bulgaria). Final Report. Polish Scientific Publishers, Warsaw.

Lejeune, M., 1987. L'art mobilier paléolithique et mésolithique en Belgique. Centre d'Études et de Documentation Archéologiques (Treignes).

Leroi-Gourhan, A., 1961. Les fouilles d'Arcy-sur-Cure. Gallia Préhistoire 4, 3–16.

Leroi-Gourhan, A., 1964. Les religions de la Préhistoire. Presses Universitaires de France, Paris.

Leroi-Gourhan, A., Leroi-Gourhan, A., 1965. Chronologie des grottes d'Arcy-sur-Cure. Gallia Préhistoire 7, 1–64.

Lévêque, F., Vandermeersch, B., 1980. Découverte de restes humains dans un niveau castelperronien à Saint-Césaire (Charente-Maritime). Cr Acad. Sci. A 291D, 187–189.

Mania, D., 2002. Der mittelpaläolithische lagerplatz am ascherslebener see bei königsaue (Nordharzvorland). Praehistoria Thuringica 8, 16–75.

Mazza, P.A., Martini, F., Sala, B., Magi, M., Colombini, M.P., Giachi, G., Landucci, F., Lemorini, C., Modugno, F., Ribechini, E., 2006. A new Palaeolithic discovery: tar-hafted stone tools in a European Mid-Pleistocene bone-bearing bed. J. Archaeol. Sci. 33, 1310–1318.

McBrearty, S., Brooks, A., 2000. The revolution that wasn't: a new interpretation of the origin of modern human behavior. J. Hum. Evol. 39, 453–563.

McDermott, F., Grün, R., Stringer, C.B., Hawkesworth, C.J., 1993. Mass spectrometric U-series dates for Israeli Neanderthal/early modern hominid sites. Nature 363, 252–254.

Mellars, A., 1996. The Neanderthal Legacy. Princeton University Press, Princeton.

Mellars, A., 1999. The Neanderthal problem continued. Curr. Anthropol. 40, 341–350.

Mellars, A., 2005. The impossible coincidence. a single-species model for the origins of modern human behavior in Europe. Evol. Anthropol. 14, 12–27.

Mellars, A., 2010. Neanderthal symbolism and ornament manufacture: the bursting of a bubble? Proc. Natl. Acad. Sci. U S A 107, 20147–20148.

Moreau, L., 2003. Les éléments de parure au Paléolithique supérieur en Belgique. L'anthropologie 107, 603–614.

Otte, M., 1979. Le Paléolithique Supérieur Ancien En Belgique. Musées Royaux d'Art et d'Histoire, Bruxelles.

Palma di Cesnola, A., 1993. Il Paleolitico Superiore in Italia. Garlatti e Razzai, Firenze.

Peresani, M., Fiore, I., Gala, M., Romandini, M., Tagliacozzo, A., 2011. Late Neandertals and the intentional removal of feathers as evidenced from bird bone taphonomy at Fumane Cave 44 ky B.P., Italy. Proc. Natl. Acad. Sci. U S A 108, 3888–3893.

Pettitt, P.B., White, M., 2012. Early *Homo sapiens* in Kent's Cavern. The whats and whens of the KC4 maxilla. Curr. Arch. 262, 20–21.

Peyrony, D., 1934. La Ferrassie. Moustérien – Périgordien – Aurignacien. Préhistoire III, 1–92.

Powell, A., Shennan, S., Thomas, M.G., 2009. Late Pleistocene demography and the appearance of modern human behavior. Science 324, 1298–1301.

Rendu, W., Beauval, C., Bismuth, Th., Maureille, M., Bourguignon, L., Delfour, G., Lacrampe-Cuyaubère, F., Turq, A., 2011. New excavation of the Mousterian site of La Chapelle-aux-Saints. In: Abstracts ESHE 2011. European Society for the Study of Human Evolution, Leipzig, p. 90.

Richter, J., 2000. Social memory among late Neanderthals. In: Örschiedt, J., Weniger, G. (Eds.), Neanderthals and Modern Humans – Discussing the Transition. Central and Eastern Europe from 50.000–30.000 B. P. Neanderthal Museum, Mettmann, pp. 123–132.

Richter, J., 2002. Die ^{14}C daten aus der Sesselfelsgrotte und die Zeitstellung des Micoquicn/M.M.O. Germania 80, 1–22.

Roebroeks, W., Corbey, R., 2001. Biases and double standards in palaeoanthropology. In: Corbey, R., Roebroeks, W. (Eds.), Studying Human Origins. Disciplinary History and Epistemology. Amsterdam University Press, Amsterdam, pp. 67–76.

Rougier, H., Milota, S., Rodrigo, R., Gherase, M., Sarcină, L., Moldovan, O., Zilhão, J., Constantin, S., Franciscus, R.G., Zollikofer, C.P.E., Ponce de León, M., Trinkaus, E., 2007. Pestera cu Oase 2 and the cranial morphology of early modern Europeans. Proc. Natl. Acad. Sci. U S A 104, 1165–1170.

Roussel, M., Soressi, M., 2010. La Grande Roche de la Plématrie à Quinçay (Vienne). L'évolution du Châtelperronien revisitée. In:

Buisson-Catil, J., Primault, J. (Eds.), Préhistoire Entre Vienne et Charente – Hommes Et Sociétés Du Paléolithique. Association des Publications Chauvinoises, Villefranche-de-Rouergue, pp. 203–219.

Sandgathe, D.M., Dibble, H.L., Goldberg, P., McPherron, S.P., 2011. The Roc de Marsal Neandertal child: a reassessment of its status as a deliberate burial. J. Hum. Evol. 61, 243–253.

Schmider, B. (Ed.), 2002. L'Aurignacien de la grotte du Renne. Les fouilles d'André Leroi-Gourhan à Arcy-sur-Cure (Yonne). Gallia Préhistoire Supplément XXXIV, Paris.

Shennan, S., 2001. Demography and cultural innovation: a model and its implications for the emergence of modern human culture. Camb. Archaeol. J. 11, 5–16.

Smith, F.H., Janković, I., Karavanić, I., 2005. The assimilation model, modern human origins in Europe, and the extinction of Neandertals. Quatern Int. 137, 7–19.

Soressi, M., d'Errico, F., 2007. Pigments, gravures, parures: les comportements symboliques controversés des Néandertaliens. In: Vandermeersch, B., Maureille, B. (Eds.), Les Néandertaliens. Biologie et cultures. Éditions du CTHS, Paris, pp. 297–309.

Speth, J., 2004. News flash: negative evidence convicts Neanderthals of gross mental incompetence. World Archaeol. 36, 519–526.

Stiner, M., 1999. Palaeolithic mollusc exploitation at Riparo Mocchi (Balzi Rossi, Italy): food and ornaments from the Aurignacian through the Epigravettian. Antiquity 73, 735–754.

Stringer, C., 1998. Chronological and biogeographic perspectives on later human evolution. In: Akazawa, T., Aoki, K., Bar-Yosef, O. (Eds.), Neandertals and Modern Humans in Western Asia. Plenum Press, New York, pp. 29–37.

Stringer, C., Andrews, P., 1988. Genetics and the fossil evidence for the origin of modern humans. Science 239, 1263–1268.

Stringer, C., Gamble, C., 1993. In: Search of the Neanderthals. Thames and Hudson, London.

Taborin, Y., 2002. Les objets de parure et les curiosa. In: Schmider, B. (Ed.), L'Aurignacien de la grotte du Renne. Les fouilles d'André Leroi-Gourhan à Arcy-sur-Cure (Yonne). Gallia Préhistoire Supplément XXXIV, Paris, pp. 251–256.

Texier, P.-J., Porraz, G., Parkington, J., Rigaud, J.-Ph., Poggenpoel, C., Miller, C., Tribolo, C., Cartwright, C., Coudenneau, A., Klein, R., Steele, T., Verna, C., 2010. A Howiesons Poort tradition of engraving ostrich eggshell containers dated to 60,000 years ago at Diepkloof Rock Shelter, South Africa. Proc. Natl. Acad. Sci. USA 107, 6180–6185.

Trinkaus, E., 2005. Early modern humans. Annu. Rev. Anthropol. 34, 207–230.

Trinkaus, E., 2007. European early modern humans and the fate of the Neandertals. Proc. Natl. Acad. Sci. U S A 104, 7367–7372.

Trinkaus, E., Moldovan, O., Milota, S., Bîlgăr, A., Sarcina, L., Athreya, S., Bailey, S.E., Rodrigo, R., Mircea, G., Higham, Th., Bronk Ramsey, C.H., Plicht, J.V.D., 2003. An early modern human from the Pestera cu Oase, Romania. Proc. Natl. Acad. Sci. U S A 100, 11231–11236.

Trinkaus, E., Shipman, P., 1994. The Neandertals. Of Skeletons, Scientists, and Scandal. Vintage Books, New York.

Vanhaeren, M., d'Errico, F., Stringer, C., James, S.L., Todd, J.A., Mienis, H.K., 2006. Middle Paleolithic shell beads in Israel and Algeria. Science 312, 1785–1787.

Watts, I., 2010. The pigments from pinnacle point cave 13B, Western Cape, South Africa. J. Hum. Evol. 59, 392–411.

Wendt, W.E., 1974. "Art mobilier"aus der Apollo 11-Grotte in Südwest-Afrika. Die ältesten datierten Kunstwerke Afrikas. Acta praehistorica et archaeologica 5, 1–42.

Wendt, W.E., 1976. 'Art Mobilier' from the Apollo 11 Cave, South West Africa: Africa's oldest dated works of art. S. Afr. Archaeol. Bull. 31, 5–11.

White, R., 1982. Rethinking the middle/upper Paleolithic transition. Curr. Anthropol. 23, 169–192.

White, R., 2001. Personal ornaments from the grotte du renne at arcy-sur-cure. Athena Rev. 2, 41–46.

White, R., 2002. Observations technologiques sur les objets de parure. In: Schmider, B. (Ed.), L'Aurignacien de la Grotte du Renne. Les fouilles d'André Leroi-Gourhan à Arcy-sur-Cure (Yonne). Gallia Préhistoire Supplément XXXIV, Paris, pp. 257–266.

Wolpoff, M., Caspari, R., 1996. Why aren't Neandertals modern humans? In: Bar-Yosef, O., Cavalli-Sforza, L., March, R., Piperno, M. (Eds.), The Lower and Middle Palaeolithic. Abaco, Forlì, pp. 133–156.

Yellen, J., 1996. Behavioural and taphonomic patterning at Katanda 9: a middle stone age site, Kivu Province, Zaire. J. Archaeol. Sci. 23, 915–932.

Yellen, J., 1998. Barbed bone points: tradition and continuity in Saharan and Sub-Saharan Africa. Afr. Archaeol. Rev. 15, 173–198.

Zilhão, J., 2001. Anatomically Archaic, Behaviorally Modern: The Last Neanderthals and their Destiny. Stichting Nederlands Museum voor Anthropologie en Praehistoriae, Amsterdam.

Zilhão, J., 2006. Neandertals and moderns mixed, and it matters. Evol. Anthropol. 15, 183–195.

Zilhão, J., 2007. The emergence of ornaments and art: an archaeological perspective on the origins of behavioural "modernity". J. Archaeol. Res. 15, 1–54.

Zilhão, J., 2011. Aliens from outer time? Why the "human revolution" is wrong, and where do we go from here? In: Condemi, S., Weniger, G.-C. (Eds.), Continuity and Discontinuity in the Peopling of Europe. One Hundred Fifty Years of Neanderthal Study. Springer, New York, pp. 331–366.

Zilhão, J., Angelucci, D., Badal-García, E., d'Errico, F., Daniel, F., Dayet, L., Douka, K., Higham, T.F.G., Martínez-Sánchez, M.J., Montes-Bernárdez, R., Murcia-Mascarós, S., Pérez-Sirvent, C., Roldán-García, C., Vanhaeren, M., Villaverde, V., Wood, R., Zapata, J., 2010a. Symbolic use of marine shells and mineral pigments by Iberian Neandertals. Proc. Natl. Acad. Sci. U S A 107 (3), 1023–1028.

Zilhão, J., Angelucci, D., Badal, E., Lucena, A., Martín, I., Martínez, S., Villaverde, V., Zapata, J., 2010b. Dos abrigos del Paleolítico superior en Rambla Perea (Mula, Murcia). In: Mangado X. (Ed.), El Paleolítico superior peninsular. Novedades del siglo XXI. Universidad de Barcelona, Barcelona, pp. 97–108.

Zilhão, J., d'Errico, F., 1999. The chronology and taphonomy of the earliest Aurignacian and its implications for the understanding of Neanderthal extinction. J. World Prehist. 13, 1–68.

Zilhão, J., d'Errico, F., 2003. An Aurignacian "Garden of Eden" in southern Germany? An alternative interpretation of the Geissenklösterle and a critique of the Kulturpumpe model. Paléo 15, 69–86.

Chapter 5

Invention, Reinvention and Innovation: The Makings of Oldowan Lithic Technology

Erella Hovers

Institute of Archaeology, The Hebrew University of Jerusalem, Mt. Scopus Jerusalem 91905, Israel, hovers@mscc.huji.ac.il

5.1. INTRODUCTION

The two most creative processes on our planet are biological and cultural evolution. It can be argued that both are Darwinian processes, regulated by blind variation and selective retention (Simonton, 1999, 2000). This statement requires qualification; unlike biological evolution, cultural evolution is goal-directed through creative individuals who struggle to construct their niche by manipulating ecological obstacles and social relations (Laland et al., 2000; Simonton, 1999; Ziman, 2000). Creativity lies at the root of the cultural diversity of modern humans. Intimately linked with notions of progress and improvement, it propels much of the dynamics of change and diversity in major cultural undertakings of contemporary cultures such as science, art, design or engineering. In the modern world, creativity often culminates in material things or ideas about how to make them (e.g., Basalla, 1988; Gold, 2007). It stands to reason that archaeologists, who work with a 2.5-Ma record of material culture, will concern themselves with identifying and explaining a great amount of cultural diversity through objects, and by implication will have a vested interest in the processes of creative thought (Mithen, 1998a).

However, the practicalities of archaeology are not conducive to studying creativity. First, the novel ideas of individuals, which are at the core of creativity, are inaccessible to archaeologists. The psychological and philosophical frameworks employed to study contemporary creativity are therefore not directly relevant for researching this phenomenon in the past (Lake, 1998; Mithen, 1998a). Second, because the archaeological study of human prehistory encompasses the dimension of time, time averaging is inevitable in all but highly exceptional circumstances of prehistoric investigations. This in turn would mask by default unique episode of creativity. Finally, the evolutionary concepts that permeate prehistoric archaeology, as well as the taphonomic constraints of the record, render individual acts invisible to us almost by definition (Bailey, 2007; Malinsky-Buller et al., 2011). Thus, a creative act, an invention that is expressed as an entirely novel type of artefact or technique, may not be recognised because it is likely to be swamped within the normal range of variation of the pertinent archaeological record. Furthermore, if a technological novelty – a new artefact shape, habitual use of a formerly unexploited raw material or the application of a new technological process – anticipates later procedures or forms, it might be dismissed as an intrusion of more recent material into older stratigraphic units (Hovers et al., 1997; Kuhn and Stiner, 1998).

Even if recognised, a creative novelty presents us with an epistemological problem, which palaeoanthropology shares with other historical sciences that are interested in unique events (Gould, 1986). In a classic catch-22 scenario, novel behaviours that cannot plausibly be accounted for in terms of prior behaviours of a species speak of cognitive capacities that can lead to innovations. Yet the very novelty of such behaviours and their uniqueness are one-time events. From the narrow viewpoint of formal ("hard") science, these first, unique events are anecdotes that do not lend themselves to proper scientific enquiry. On the other hand, behaviours that are scientifically useful – that is, which became normative and were often repeated, hence recognised in the fossil (in our case, the archaeological) record – speak of conditioned responses, contact and imitativeness (Dennett, 1997). They can tell us very little about creativity. For these reasons, researchers of the Palaeolithic have reasonably abandoned attempts to pin down elusive cognitive inventions, i.e., material culture "firsts" that express creative potential, in the records they study. Rather than looking at single artefacts or types, they have opted to look at the rates of turnover and degrees of variety in artefact forms (Kuhn and

Stiner, 1998: 146) which represent *realised* creativity – that part of the creative potential that became "innovations", stabilised in the archaeological record through social and demographic processes (van der Leeuw and Torrence, 1989; Hovers and Belfer-Cohen, 2006; Kuhn and Stiner, 1998; Mithen, 1998b; Renfrew, 1978; Shennan, 2001).

The Oldowan, being the first known recognisable technocomplex of hominin material culture, appears at face value to be "a natural" for investigating creativity in early hominins. While clearly not on a par with the creativity that we observe among humans in the modern world, the Oldowan appears to be "something" that emerged out of "nothing" (Rogers and Semaw, 2009). Current evidence tells us that its first occurrence was in a well-defined region in Ethiopia, where few sites are known within a temporal span of 2.6–2.5 Ma (Semaw, 2000; Semaw et al., 2003). The localities represent short occupation periods, probably on the order of tens of years, or less, in each case.

A feature that makes the Oldowan a recognisable archaeological entity is the existence of spatial clusters of anthropogenic artefacts, referred to as "sites" or "occurrences", for the first time in the hominin record. As discussed below, this phenomenon does not necessarily pertain to creativity. Another conspicuous novelty that defines the Oldowan is the habitual flaking of stone. The early sites mark the onset of roughly a million years (2.6–1.5 Ma) of Oldowan stone toolmaking. Broadly associated with cutting activities (Toth, 1985; Toth and Schick, 2009), the specific functions of Oldowan stone tools are not clear. Regardless, in ecological terms stone toolmaking appears to have been a beneficial adaptation that played an important role in constructing the hominin niche (e.g., Ambrose, 2001; Braun and Hovers, 2009; Laland and O'Brien, 2010; Laland et al., 2000). The technological invention of making stone artefacts spiralled into a widespread innovation (as defined and analyzed by Renfrew, 1978; Spratt, 1982) within a time span of a few hundred thousand years.

This paper addresses two questions about the Oldowan. First, was the Oldowan a creative event? Are there elements of the Oldowan that may be legitimately considered as the outcome of creative acts? Second, was the spread of the Oldowan after 2.5 Ma due to social learning and cultural transmission as opposed to independent reinventions which would be considered, by definition, creative?

Accepting that cognitive abilities have evolved throughout prehistory, the skills and capacities of extant humans are too distant evolutionarily from those of our Oldowan-making ancestors to serve as direct analogues of the abilities of early hominins. On the other hand, because of the close evolutionary relationship between panins and hominins, creative abilities of extant chimpanzees (and to a lesser degree, of other nonhuman primates) reasonably outline the lower boundary of such capacities in early hominins. Insights into creative cognition in early hominins might be sought after in the range defined by creative mechanisms in extant humans on the one hand, and of tool-using, tool-making nonhuman primates on the other. Importantly, while cognitive characteristics may possibly be identified in extant primates, the drawing of inferences concerning extinct hominin taxa must rely on material objects. For an element of material culture to be considered as evidence for early hominin creativity, it would have to be derived ("autapomorphic") and different from the material culture behaviours of chimps, rather than shared ("synapomorphic") by hominins with one or more nonhuman primate.

Did the Oldowan technocomplex spread as a series of "reinventions"? At first glance, this is not a likely scenario, because the probability of independent "reinvention" of the same set of technological behaviours would be low. It is also unlikely that such "reinventions" would occur independently at more or less the same time. Epistemological biases, too, may lead to rejection of the possibility of recurrence. As archaeologists, we are more accustomed to frame models of accretional changes through human evolution (e.g., Gamble, 2007). Yet, a growing body of theoretical work and archaeological case studies (e.g., Belfer-Cohen and Hovers, 2010; Enquist et al., 2008; Hovers, 2009a; Hovers and Belfer-Cohen, 2006; Kuhn and Hovers, 2006; Powell et al., 2009) has shown that such a linear framework should not be applied uncritically to all instances of culture change. Given these shifts in worldview, the notion of a gradual spread and growth of Oldowan technology is not necessarily a fact of the archaeological record, but a working hypothesis that needs to be considered critically.

In view of the caveats reviewed above, it would be foolhardy to attempt clear-cut answers to any of the issues raised in a discussion of Oldowan creativity. Rather, my goal is to present, and to some degree explore, a range of feasible scenarios (and not a comprehensive one at that). To address the two questions that are the focus of this paper, it is necessary to first define elements of creativity as discussed in philosophical and psychological literature on creativity among modern humans. Isolating these elements would facilitate exploring their presence in the archaeological record. I will present a brief overview of some relevant thinking on these issues, emphasising insights that are more pertinent to the issues discussed in this paper. This is followed by an overview of the Oldowan archaeological record, emphasising the earlier part of its million-year time span. Finally, material culture behaviours of nonhuman primates and their creativity-related interpretations are brought to bear on the discussion in this paper.

5.2. WHAT DOES IT TAKE TO BE CREATIVE?

Creativity is said to entail the production of some entity that is simultaneously both original and adaptive (Simonton,

1999). For most of us, the term "creativity" denotes novel ideas or surprising combinations of old ideas that impress and surprise us because of their improbability within our familiar conceptual frameworks (Boden, 1998). The term typically connotes a rare cognitive ability that graces a small number of exceptional individuals. We would not bother to qualify this ability with any adjective in order to explain that we refer to a rare gift. Psychologists, however, distinguish between "exceptional" and "mundane" creativity.

An idea would be considered a radical novelty – an act of exceptional creative thinking – if the individual who had it could not have had it before and if nobody else, in all of human history, has ever had it before (Boden, 1998). This is a philosophical distinction. Psychologically, the ascription of creativity always involves an explicit reference to some generative system and a particular set of generative principles, with respect to which the genuinely original or radically creative idea steps out of bounds. For such transformation, analogical thinking is required. This is the ability to enlist and cross-reference information from a variety of cognitive domains to bear on a problem in yet another domain (Boden, 1998; Mithen, 1996). This in turn may lead to an inflation of novel associations and ideas that are mostly meaningless (Perkins, 2000). For exceptional creativity to emerge from analogical thinking, several cognitive mechanisms must be employed to curb the combinatorial explosion. By definition then, it is the existence of conceptual constraints that makes exceptional creativity possible, since random processes alone can result only in first-time curiosities (Boden, 1998; see also Dennett, 1997). True exceptional creativity stems from negation of a conceptual constraint and/or transformations of conceptual spaces.

The other category of creativity – mundane creativity – pertains to the generative potential inherent in the operating characteristics of most human brains (Barsalou and Prinz, 2000). An idea that an individual has for the first time is perceived as creative, regardless of how many times it might have already occurred to other people. These types of novel ideas can be described and/or produced by the same sets of cognitive generative rules as other, already familiar ideas (Boden, 1998), and do not amount to earth-shattering novelties.

Creativity plays a role in goal-directed cultural evolution. On the level of the individual, the creative process is Darwinian, because whenever a problem (practical or intellectual) requires genuine creativity, there will come a point where the individual falls back on a blind-variation process (playful exploration, free association or haphazard tinkering with objects) (Campbell, 1960; Simonton, 1999). In the interval between an individual's first attempt to solve a problem and reaching a decision about a preferred solution, the person is exposed to a largely random influx of stimuli that are constantly interacting, creating cognitive networks related to the problem. This produces a series of alternative formulations, some more fruitful than others but only one of which will eventually lead the individual to a solution. The order in which the new conceptions appear is random in relation to the problem. This means that the mind of the creative individual is engaged in an inadvertent blind-variation process (Simonton, 1999) that is not necessarily restricted by the normative options recognised by the social or cultural network within which that individual operates. A creative act – the production of a novelty – is a consequence of searches within a conceptual space of possibilities (i.e., possible inventions, theories and associations) that is formed by these variations. This space is not featureless. It has a "topography" that can make solutions either relatively accessible or very hard to find (Perkins, 1992, 2000).

Such modelling of the "possibilities space" helps explain the differences in the working mechanisms and the products of exceptional and mundane creativity. A "wilderness" type of space is when suitable solutions are sparsely scattered in a large space of adequate but not perfect solutions. Consider a writer searching for the perfect adjective among a number of nearly synonymous words or think about an engineer who is looking for the exact right material for building an object, out of many with almost identical qualities. Another conceptual space is modelled as a "plateau", where large regions are homogeneous. Possibilities differ very marginally and there is no clear indication as to which possibility is preferable to others. Then again, the space of possibilities may be modelled as a "canyon" – the search for the right solution takes place within well-defined and confined boundaries while that solution may occur in a different region of the possibility space. A fourth type of possibility space is modelled as an "oasis" where promising but only partial solutions can be found. Although they cannot be tweaked into becoming the right solutions to a problem, the search remains confined to this area because of the partial promise offered by the existing possibilities. Using the analogy of searching for gold in the Klondike, Perkins (1992) refers to the tough possibility spaces formed by these topographies as *Klondike spaces*.

The difficulties imposed by some or all of the topographies of the Klondike space constitute the limitations that are requisite for exceptional creativity. The right solution may be hard to find because it is difficult to identify, as in the wilderness and/or plateau models, or because it requires resisting familiar near-good solutions, e.g., solutions that conform with known social/cultural norms (the oasis model) or because finding the right solution requires reconfiguring the problem from scratch by moving to an entirely different part of the conceptual space (the canyon model). The creativity of artists such as Shakespeare, Picasso or Pollock, or of scientists such as

Einstein, is of this type, falling outside of the proverbial box – i.e., the options offered by familiar norms of the society/art or paradigms of science within which one operates.

In contrast, mundane creativity relies on more routine problem-solving processes. A person must go beyond the basic information in order to come up with a solution to a given problem, but in this case the search takes place within a *Homing space* (Perkins, 1992), a possibility space where there is a clear gradient of options that can be followed to the right solution (e.g., social or scientific norms) through an iterative and organised search, for example, the processes of hypothesis testing in positivist sciences. This guarantees finding reliable solutions within a familiar conceptual box.

The two types of creativity represent endpoints on a continuum of human creativity (Ward et al., 2000: 4-5). They are also relational in that there are factors besides the logic of the problem that determine the topography of the conceptual possibility space for a given problem. The topography of the possibility space depends on the way a problem is encoded by the cognitive agent. Consider stone-knapping skill. If a person is a novice to a certain problem, his search for solutions will be conducted in Klondike spaces. An experienced person would conduct the same search in a homing space; or consider the social context of conducting a search. A robust system of knowledge transmission is more likely to yield searchers within homing spaces, whereas a less robust system will not offer beaten tracks and will force searches in *Klondike spaces*.

The creative process (i.e., the formation of blind variation, the search for solution and the solutions that a problem solver comes up with) requires minimally analogical thinking that links conceptual spaces ("cognitive domains"; e.g., Gardner, 1983; Mithen, 1996), creates novel relationships between phenomena and/or objects in the world and leads to new perceptions, thoughts and actions. It has also been suggested that possession of theory of mind (ToM) – the capacity to understand conspecifics as intentional agents like the self (Tomasello, 1999: 53) – is essential for the exploration and transformation of conceptual spaces (Mithen, 1998b).

These insights into the creative process are based on the working of the modern human mind. Still, using them to lay out some fundamental mechanisms of the creative act allows us to explore the Oldowan archaeological record for evidence that may bear on the role of creativity in early hominin material cultural evolution. If a real break from a behavioural repertoire that hominins share with apes can be shown, an argument can be made for exceptional creativity. Conversely, it would be reasonable to suggest that novelties of the Oldowan archaeological record represent mundane creativity if a given behaviour is similar, shared with or is a direct expansion of behaviours known among apes.

5.3. A BRIEF OVERVIEW OF THE OLDOWAN

The earliest known hominin-made lithic assemblages, consisting of stones that have been deliberately flaked through percussive blows, appear in archaeological localities within a restricted region in Gona (the Afar depression, northeast Ethiopia) at 2.6 Ma. The geographic distribution and number of localities with similar lithic assemblages increased during 2.4–2.3 Ma, when they were found in other regions of Ethiopia and Kenya. Another geographic expansion and quantitative increase in the number of localities are noted in the period 2.1–1.9 Ma, with sites located in the Rift Valley of East Africa and in sedimentary basins in North and Central Africa. Contemporaneous sites in South Africa are in secondary context (Blumenschine et al., 2008; Braun and Hovers, 2009; Chavaillon, 1976; Chavaillon and Piperno, 2004; Delagnes et al., 2011; Feibel et al., 1989; Harris et al., 1987; Hovers et al., 2008; Howell et al., 1987; Isaac and Isaac, 1997; Kimbel et al., 1996; Leakey, 1971; Plummer, 2004; Plummer et al., 1999; Roche et al., 2003; Rogers and Semaw, 2009; Sahnouni, 2006; Sahnouni et al., 2002, 2011; Semaw et al., 2003; Stout et al., 2010; Texier, 1995).

During the period 2.5–2.0 Ma at least four species of hominins (*Australopithecus garhi*, *Paranthropus boisei*, *Australopithecus aethiopicus*, and *Homo* sp.) are known from East Africa (Wood, 1997) and one (*A. africanus*) from South Africa. Between 2.0 and 1.0 Ma, *A. aethiopicus* disappears from the East African record, while *H. rudolfensis*, *H. habilis* and *H. erectus* appear for the first time. The latter is also the hominin that first appears out of Africa associated with Oldowan lithic technology nearly 2.0 Ma (Ferring et al., 2011; Rightmire et al., 2006). In South Africa, *A. robustus* and *A. sediba* (Dirks et al., 2010; Pickering et al., 2011) appear in the fossil record during this time. The latter is argued to be potentially a stone toolmaker, based on its hand morphology (Kivell et al., 2011).

Starting ~2.35 Ma, fossil remains that belong to early *Homo* were found in stratigraphic units that contain artefacts (Prat et al., 2005) or eroding directly from artefact-bearing sediments (Kimbel et al., 1996), the latter providing the strongest empirical association of early Oldowan tools and their plausible makers. *A. garhi* (loosely associated with cut-marked bones at ~2.5 Ma) was suggested to be an early non-*Homo* stone toolmaker, but it was found some 60 km from the Gona sites that yielded stone tools of this age (de Heinzlein et al., 1999). In Koobi Fora and Olduvai, 2.0–1.5 Ma-old remains assigned to robust australopiths were found near lithic artefacts but without secure associations (e.g., Isaac and Harris, 1978; Wood, 1991). Based on this fossil record, it cannot be ruled out that more than one of the hominin forms that existed during 2.5–1.0 Ma could have been involved in stone-tool manufacture and use. At the same time, it is apparent that members of the genus *Homo* demonstrate a significant brain expansion during the period of 2.0–1.0 Ma, which may be correlated with a tool-making emphasis in their evolution and possibly with the ability to expand out of Africa, using the simple core-and-flake technology that constitutes the earliest material culture record (Toth and Schick, 2009).

Pollen, fauna, sedimentology and stable isotope data suggest that in East Africa, where the majority of primary-context sites are located, hominins exploited a variety of savanna habitats, ranging from patches of open grasslands to riparian woodlands. The Oldowan sites are frequently found in past well-watered areas, possibly wooded, within fluvio-lacustrine systems (Ashley et al., 2010; Hovers et al., 2008; Howell et al., 1987; Kimbel et al., 1996; Levin et al., 2004 and references therein; Reed and Geraads, in press). To date, Kanjera south is the only known Oldowan site located in truly open grassland (Plummer et al., 2009).

The Oldowan record consists almost entirely of nonperishable stone tools. In addition, a small number of polished and striated bones have been reported and interpreted as possible digging sticks or termite fishing tools (Backwell and d'Errico, 2003; d'Errico et al., 2001), suggesting that toolkits of early hominins may have been more variable than seen from a partial archaeological record. The identification of places on the landscape as Oldowan sites is based on the spatial clustering of artefacts. Spatial associations of lithic artefacts with fossil animal bones, mostly of medium-to-large-size animals, have been observed in few occurrences, mostly postdating 2.0 Ma.

Currently, the term "Oldowan" has come to formally denote a technocomplex that includes a few material culture expressions. Its more colloquial use is in relation to lithic assemblages, based on the historical precedence of discoveries in Olduvai Gorge in the mid-20th century (Leakey, 1971). Early Oldowan stone artefacts are primarily products of hand-held percussion, where the knapper holds a stone hammer (typically a cobble) in one hand and uses it to hit a stone held in the other hand at an oblique angle. This operation results in two types of objects – the detached flakes and the piece from which they were removed, now defined as "core" ("core tool" in Leakey's (1971) typology). The use of bipolar flaking, whereby a core is placed on an anvil and struck vertically with a hammer, is less common and is related mainly to small size and lesser quality of raw material used.

Raw material selectivity is part of the Oldowan lithic technological behaviour. Various combinations of selection criteria (e.g., preferential exploitation of angular cobble morphologies that facilitated the initiation of the flaking process, nuanced preferences for grain-size and homogeneous rocks) were applied to facilitate the flaking process and to obtain edge durability during use (Braun et al., 2009; Harmand, 2009; Stout et al., 2005, 2010). In the early Oldowan, raw material acquisition was local and transport distances from source areas to the locations that we identify as archaeological occurrences were small (Harris and Capaldo, 1993; Rogers et al., 1994; Goldman-Neuman and Hovers, 2009). After ca. 2.0 Ma, raw material transport distances increase and selection criteria seem to be more complex. A notable change is a growing tendency to link certain raw materials to the production of certain tool forms (see Goldman-Neuman and Hovers, 2012; for a recent discussion). With the geographic expansion of Oldowan occurrences during this time, subtle, nondirectional shifts in stone-toolmaking practices may have led to the emergence of localised technological variants (Stout et al., 2010).

Leakey's (1971) relatively elaborate typology of Oldowan artefacts was based mainly on the varied morphologies of the core tools and retouched artefacts in sites postdating 2.0 Ma, but such variation is hardly known from the earlier Oldowan occurrences. Moreover, some of these classifications correspond to morphologies that are currently known to have formed by depositional and taphonomic processes rather than hominin preferences for specific shapes during knapping (e.g., de la Torre and Mora, 2005; de la Torre and Mora Torcal, 2005). In other cases, the redundant shapes of lithic items could be inadvertent by-products of percussive flaking (Sahnouni et al., 1997), similar to the emergence of types from chimpanzee nut cracking (e.g., the pitting of anvil stones as a result of repeated use; Carvalho et al., 2008). Artefacts that underwent secondary modification by retouch might be an important exception to this rule (Kitahara-Frisch, 1993), but again such items are extremely rare in the earliest Oldowan assemblages. Thus, the earliest Oldowan is perceived as essentially a core-and-flake lithic industry.

The use of Oldowan stone tools involved two basic modes of actions upon nonlithic materials, roughly corresponding to the two artefact classes formed during knapping. Flakes provided sharp edges, presumably for the purpose of cutting animal tissue (in the process of butchery or skinning) and plant materials. The sharp edges of Oldowan tools were once thought to be part of an adaptive complex related specifically to the appearance of hominin carnivory (Isaac and Crader, 1981). In effect, because of the near-absence of robust use-wear studies (though see Gibbons, 2009; Keeley and Toth, 1981), the functions of Oldowan lithic artefacts have been deduced from a small number of experimental studies (e.g., Braun et al., 2008; Jones, 1994; Toth, 1985). Identified activities include cutting soft plants and scraping wood (Keeley and Toth,

1981), possibly indicating the shaping of wood or bark into digging sticks and spears, for which there is now sporadic evidence among wild chimpanzees (Hernandez-Aguilar et al., 2007; Pruetz and Bertolani, 2007). Similarly, sharp-edged core forms could also be used for a range of chopping and scraping activities (de la Torre and Mora, 2005; de la Torre and Mora Torcal, 2005). The cores themselves, as well as unflaked stones, may have been used as hammers to pound, rather than cut, a range of organic materials, again similar to the behaviour seen among chimps (de Beaune, 2004; Toth, 1985, 1997; Toth and Schick, 2009).

5.4. "SOMETHING OLD, SOMETHING NEW"

The Oldowan appears to mark significant changes in hominin behaviours compared to what is known (or rather, unknown) from the preceding hominin record. Hypotheses that explain the behavioural changes run the gamut from ecological adaptations in response to environmental changes, to major shifts in social and economic organisation, including a shift to food sharing, change in settlement patterns, and increasing reliance on meat (for a recent review, see de la Torre and Mora, 2009). When we ask whether these are true innovations, epistemological caveats immediately come to mind. First among these is the fact that the available record is distorted by differential preservation biases. Within this incomplete record, the recognisable archaeological signatures that stand out are the occurrence of identifiable clusters of lithic items – with or without faunal residues – and the occurrence of the first recognisable systematic stone flaking. For each of these archaeological signatures, we must ask whether they are indeed innovative.

5.4.1. The Occurrence of Spatial Clusters

Because natural processes can lead to the occurrence of Oldowan-like artefacts (e.g., Clark, 1958), it may be difficult to identify the formative agent when only isolated stone items are found. It is "…the localised accumulation of refuse which is what has made the archaeological study of prehistoric life possible" (Isaac, 1971: 279) and allows the distinction between "sites" and low-density background distributions ("scatters") of artefacts and/or bones modified by hominins (Isaac, 1981; Isaac and Harris, 1980). The definition of the Oldowan rests on these clusters, which appear for the first time in the record – but do they constitute a creative novelty?

Initially, the answer was in the affirmative. Oldowan archaeological occurrences were thought to represent unique land-use patterns and unique human social practices involving division of labour, food sharing and occupation of home bases (Isaac, 1971, 1981). It has been argued (Potts, 1988, 1991) that the spatial clusters could have emerged if early hominins shifted their behaviour so that transport of separated stones and of subsistence materials converged on common places within their foraging range, where these resources were cached. This behaviour would render any movable resource within this range accessible for processing. However, current thought casts doubts on this interpretation. The appearance of Oldowan sites may have less to do with purposeful hominin behaviour and more to do with the vagaries of time and taphonomy (e.g., Blumenschine, 1991). The caching hypothesis (Potts, 1991) was offered in explanation of the *emergence* of the Oldowan at ~1.9 Ma or slightly later. Yet, known Oldowan sites that predate 2.0 Ma are found mainly near their raw material sources, while artefacts or cores were moved into these localities and taken out (Delagnes and Roche, 2005; Goldman-Neuman and Hovers, 2012; Hovers et al., 2008; Semaw et al., 2003; Toth et al., 2006). In view of this, these early sites are not likely to have been caching localities.

In many cases, the spatial associations of lithics and bones may be due to the palimpsest nature of the early sites, resulting from several discrete and unrelated episodes of accumulation (and removal) by hominin, faunal and/or nonbiological agents (e.g., Binford, 1981; Bunn et al., 2010; Domínguez-Rodrigo, 2009; Egeland et al., 2004; Isaac, 1978, 1983, 1984; Petraglia and Potts, 1994; Schick, 1987). Palimpsests in turn suggest a nonrandom use of foraging territories, where specific locations are revisited preferentially. This behaviour is not unlike that of wild chimpanzees, which show clear preferences for specific topographic configurations, vegetation and seasonal changes in plant, water and shelter availability. Certain locations are reused more often than others, and over time "sites" are formed (Carvalho et al., 2008; Hernandez-Aguilar, 2009; Sept, 1992). Typically, such sites are not preserved for a long time, because the impact on the nesting trees, or the debris of tools made of plant materials, does not survive. However, where chimpanzees use stone hammers to pound nuts, the hammers and their unintentional stone debris survive, and create visible clusters – sometimes of considerable antiquity – that superficially mimic Oldowan sites (e.g., Carvalho et al., 2008; Mercader et al., 2002, 2007).

The specific land-use patterns and associated behaviours of early stone toolmakers cannot be considered novel solutions to ecological demands or to the requirements of a novel social organisation. Rather, the phenomenon of "sites" may be a by-product of the fact that hominins, unlike nonhuman primates, engaged intensively in activities that involved durable raw materials.

5.4.2. Broadening of the Dietary Niche?

It is generally accepted that vegetable foods played a major role in the hominin dietary niche, although direct evidence for this is lacking. The use of plants for food as well as tools is assumed on the basis of analogies with nonhuman primates, ecological reconstructions of hominin environments, isotopic analyses and studies of tooth wear in early hominin taxa (e.g., Cerling et al., 2011; Copeland, 2009; Hernandez-Aguilar et al., 2007; Lee-Thorp et al., 2010; Sept, 1986, 1994; Ungar et al., 2006, in press; van der Merwe et al., 2008; Wrangham et al., 2009). Dietary choices of hominins during the time span of the Oldowan were rather flexible. Even the robust australopiths, once considered as dietary specialists, are now known to have subsisted on a large variety of plants (Ungar et al., 2006, in press), as hypothesised by Isaac (1971: 279). The difference between the two genera may have been in their choice of fall-back foods rather than year-round dietary preferences, with *Homo* relying on meat and *Paranthropus* turned on abrasive and hard plant foods (Alemseged and Bobe, 2009). This evidence is consistent with the variable habitats documented for the early sites.

Current evidence challenges the existence of functional relationship between artefacts and faunal remains in spatial clusters dating to the early Oldowan. At best, such a relationship is equivocal (Domínguez-Rodrigo and Martínez-Navarro, 2012). Evidence for percussion, pitting or fracturing of bone for access to marrow is scarce in pre-2.0-Ma sites. Securely identified cut marks associated with early Oldowan sites in Gona occur mainly as surface finds (Domínguez-Rodrigo et al., 2005, 2007), whereas other isolated instances of cut marks from 2.5-Ma contexts, reported from the Middle Awash, are not associated with lithic artefacts (de Heinzlein et al., 1999).

Actualistic studies were undertaken in order to understand the taphonomic scenarios that could potentially cause modification marks on bones in sites dating to the late Oldowan, when faunal remains are more frequently associated with lithic artefacts and bear more modification marks. The analytical results as well as their behavioural implications are controversial. Marks are sometimes interpreted as evidence that Oldowan hominins were scavengers that occupied a marginal place in the carnivore guild and had only late access to low-quality remains of carnivore kills (Blumenschine, 1987, 1995; Blumenschine et al., 1994, 2007). Alternatively, the data are interpreted as evidence of hominins' access to animal resources through confrontational scavenging or hunting (Bunn, 1986, 2007; Domínguez-Rodrigo, 2002; Domínguez-Rodrigo and Barba, 2006; Domínguez-Rodrigo et al., 2007, 2010; Plummer, 2004).

Primates, especially chimpanzees, are known to crave meat and obtain it by active hunting of small animals (e.g., Boesch and Boesch, 1989; Pruetz and Bertolani, 2007; Stanford, 1996). The taphonomic signatures of hunting chimpanzees are similar to those of feline and canine predators sometimes implicated in the bone accumulations associated with Oldowan artefacts. Scavenging is rare among chimps, and when it occurs it is usually on relatively fresh kills (Plummer and Stanford, 2000; Ragir et al., 2000; Tappen and Wrangham, 2000). It has been suggested that the main advantage conferred on nonhuman primates by meat-getting may be the enhancement of individual social and reproductive status rather than nutritional (e.g., Speth, 2010). Against this background, it is difficult to suggest that the early Oldowan presents radical diet-related inventions that differ essentially from the traditions shared by a number of primate species, i.e., hominin adaptations do not bear the mark of creative problem-solving with this regard.

There is one aspect of the Oldowan faunal record that speaks to an expansion of the range of meat-getting opportunities during the late Oldowan. Modelling of the structure and niche partitioning of East African Plio-Pleistocene predator communities suggests that hominins invaded (or rather, created for themselves) a very narrow niche between those of full predators and confrontational scavengers (Brantingham, 1998). By experimental and taphonomic accounts, scavenging of (nearly fresh) large animals might have been an addition to an older tradition of small mammal hunting that had only low visibility in the hominin record (Plummer and Stanford, 2000).

5.4.3. Systematic Stone Toolmaking

According to "older than the Oldowan" hypotheses (Mercader et al., 2002; Panger et al., 2002), the Oldowan is an extension of skills that were part of the primate behavioural package. Tool use is indeed a "plesiomorphic" trait, shared by a number of primate taxa. Its existence among wild populations depends on suitable ecological niches (especially in the context of extractive foraging). It is also related to the manipulative skills of these populations, which are some measure of the intelligence that enables rapid acquisition of complex skills through invention and learning. Social tolerance in gregarious settings facilitates invention as well as transmission of new ideas (van Schaik and Pradhan, 2003).

The majority of tools used by nonhuman primates are made of plant materials. Rocks are involved in defensive tool use (when stones are rolled from a cliff face to intimidate rivals or predators). Among chimps and capuchin monkeys, stones are also used as extractive tools, specifically as hammers and anvils in the context of nut cracking (e.g., Bania et al., 2009; Boesch and Boesch, 1990; Boesch et al., 2009; Carvalho et al., 2008; Fragaszy et al., 2004; Hernandez-Aguilar et al., 2007; McGrew, 1992; Morgan and Abwe, 2006; Moura and Lee, 2004; Pruetz and Bertolani, 2007; Sanz and Morgan, 2007, 2009; Sanz et al., 2004; van Schaik and Knott, 2001; van Schaik et al., 1999; Westergaard, 1995).

In contrast to the ubiquity of tool use among primates as a whole, toolmaking in the wild is known only among chimpanzees and (to a lesser degree) orangutans (van Schaik and Pradhan, 2003; van Schaik et al., 1999). Ape toolmaking involves raw materials made from plants (e.g., preparation of twigs for ant dipping and termite fishing) without deliberate modification of stone. Because toolmaking is exhibited by chimps and orangutans in the wild, it can be assumed to be part of a behavioural and technological package of the common ancestor of hominins, panins and pongins.

The tools of nonhuman primates fall into three major categories of complexity, which Read and van der Leeuw (2008) associate with growth in working memory capacities. The first and most basic category is of objects with functional attributes for a given task present in the object's natural state; modification is not necessary. If the cut-marked bones from 3.7-Ma deposits in Dikika, found without evidence of stone tools (McPherron et al., 2010), are accepted as legitimate evidence, it is possible that their occurrence reflects this category of tools – the use of a "ready-made" tool (in this case, a naturally broken stone). A slightly more complex type of tools used by apes involves the pairing of unmodified objects, for example, the hammer and anvil in chimpanzee nut cracking.

A more complex form of ape tools occurs when the potential functionality of an object is perceived, and then realised by making simple modifications; those may be repeated several times during a tool's use life. Raw material selectivity is built into this simple categorisation of complexity of tools, especially among chimpanzees. They are selective of the raw material type and weight of the hammers used for nut cracking (e.g., Carvalho et al., 2008; Kortland, 1986; Sakura and Matsuzawa, 1991; Sugiyama and Koman, 1979) and tend to be particular about the source, size and shape of twigs and branches that they modify into tools for ant dipping and termite fishing (e.g., McGrew, 1992, 1993). Tools or would-be tools are sometimes transported over tens of metres, from the location of raw material procurement to the place of the anticipated activity. Consistent with these observations, studies of tool use among chimpanzees and orangutans suggest that both species possess at least rudimentary episodic memory, namely, the ability to plan for future needs (Mulcahy and Call, 2006; Osvath 2009, 2010; Osvath and Osvath, 2008).

Note that beyond modification of the tool itself, there are elements of what can be called "passive modification" in the latter two types of tool use. Objects may be used in their natural form or modified, yet they are often selected from a population of similar objects to best fit the anticipated task, and may be transported to the place where they are to be used. Thus, characteristics of early Oldowan selectivity of lithic raw material and initial transport over short distances are comparable to those observed among chimps (Goldman-Neuman and Hovers, 2012) and do not necessarily correspond to a novel behaviour.

By contrast, the earliest recognised stone-knapping procedures deviate sharply from anything that is observed among nonhuman primates. Beginning ca. 2.4 Ma, *H. habilis* (and *H. erectus*) used stone tools in the context of extractive foraging, to increase the range of foods to which they would have had access (Ungar et al., in press). In this, their approach to toolmaking was very different from that of extant apes. To begin with, early hominins purposefully modified stone. To date, nonhuman primates in the wild have not been observed flaking or breaking stones in order to obtain a desirable tool. The lithic assemblages of chimpanzees result from accidental breaks of hammer stones and anvils during nut cracking (e.g., Carvalho et al., 2008; Mercader et al., 2002, 2007). There are no reports that such accidental debris has been re-evaluated and perceived as functional tools that can be used to obtain specific goals, namely cutting.

Early Oldowan knappers did not flake cores in order to obtain preconceived forms, yet they did end up with a number of redundant core morphologies. This may have been an outcome of the need to constantly reposition the concentration of mass on the core so that knapping could proceed (Moore, 2010). The procedure was carried out by an intuitive grasp of conchoidal fracture mechanics (e.g., incident angle of less than 90° when the core is hit with a hard hammer and an understanding of the way that flakes spall off a core). The identification of the geometric rules needed for flaking and their combination with the motor actions required to produce a flake constitute relatively simple "grammars of action" (Moore, 2010). In the Gona, Hadar and Lokalalei sites (2.6–2.34 Ma), cobbles were typically knapped along their short axis in order to produce more flakes from a relatively extensive striking platform (Delagnes and Roche, 2005; Hovers and Davidzon, in preparation; Semaw, 2000: fig. 6; Semaw et al., 2003: fig. 4:2, 4; Stout et al., 2010: fig. S1). Following the formation of the first striking platform, blows were aimed to establish specific spatial relationships between a given flake and its predecessors in neighbouring regions of the core, including removal of flakes to amend the core's surface geometry (Hovers, 2009b). Topological relations were actively formed and then maintained throughout the flaking process, overcoming physical constraints imposed on the process by the nature of the raw material (Read, 2008; Wynn, 1989). Flakes from consecutive platforms in the same area of the core were arranged so as to avoid hitting the core repeatedly at the same point, thus minimising accidental breakage (Fig. 5.1). This complex technological behaviour has not been seen in the wild, where it is not promoted by incentives from humans (Toth and Schick, 2009).

Anatomical limitations, specifically in hand structure and physiology and in arm musculature as well as brute

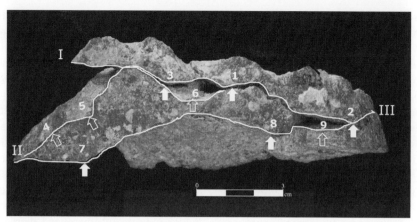

FIGURE 5.1 A set of refitted flakes from A.L. 894, Hadar, Ethiopia (∼2.35 Ma). Flakes are numbered in the order of their removal from the core. The arrows mark the point of impact of the hammer stone on the core edge ("bulb of percussion") from which flake separation propagated from the core. Empty arrows mark the points for flakes that were not refitted. The while lines correspond to consecutive striking platforms of the cores as they change during the process (I is the earliest, III the latest in the process). Note that the points of impact are not placed one underneath the other but are slightly offset, creating sets of repeated spatial relations.

strength (Marzke, 1997; Tocheri et al., 2008; Rolian et al., 2011; Toth and Schick, 2009) provide only part of the explanation for the difference in behaviours of panins and early hominins, given that the hominin hand retained some primitive features till long after the emergence of stone toolmaking (Tocheri et al., 2008). On the other hand, the cognitive underpinnings of the ability of hominins to engage in this complex toolmaking behaviour are little understood, but can perhaps be informed by those of chimps. It has been argued that the inability of chimps to engage spontaneously in hominin-type toolmaking is related to their limited short-term memory capacity, due to which they are not able to mentally retain the numerous procedures involved in the process of knapping stone tools (Davidson and McGrew, 2005; Read and van der Leeuw, 2008). Importantly, Pygmy chimpanzees (*Pan paniscus*) seem to be cognitively capable of learning from a human instructor some (but not all) of the *procedures* involved in Oldowan-style stone toolmaking, and to overcome some of the anatomical restrictions imposed by their hand structure (Schick et al., 1999; Toth et al., 1993), yet whether they comprehend the *principles* of the behaviour remains a moot point. By implication, the earliest Oldowan toolmakers were cognitively able to break through the "glass ceiling" imposed by the limitations on episodal short-term memory (Andersson, 2011; Osvath and Gardenfors, 2005). Early Oldowan toolmaking represents a true break from the capacities of nonhuman hominids.

Oldowan stone tools mark a deviation from a tradition of tool use and relatively noncomplex toolmaking, presumably shared by the ancestors of Oldowan hominins and of extant apes. Can these stone tools be considered a radical novelty? Exceptional creativity does not require that actions or objects be invented from scratch and out of the blue, but rather that familiar actions, materials and objects be linked in novel, surprising ways. The technological repertoire of chimpanzees, defining our lower boundary for inferring hominin abilities, or the tool-use behaviours shared by a number of primate species (presumably inclusive of pre-2.6-Ma hominins known from the fossil record) do not suggest that stone-tool knapping emerged within the "homing space" of routine material culture practices and traditions. In early Oldowan stone-knapping procedures, familiar elements (stone, gestures of percussion shapes of accidentally removed flakes with sharp edges) were linked in novel ways that led to a useful solution. Not only were cutting edges achieved, but they were obtained in a relatively large number from a given core. The first instances of Oldowan stone toolmaking should be perceived as the outcome of an act of exceptional creativity.

5.4.4. The Later Oldowan: Re-invention and Cultural Transmission

The Oldowan technocomplex is defined by controlled application of conchoidal fracture mechanics to produce sharp edges, using very basic "grammars of action" (Isaac, 1976; Moore, 2010; Toth, 1985). Stout et al. (2010) proposed that the Oldowan consisted of nondirectional technological variability without substantial changes for nearly 1 million years. It has been suggested that after 2.5 Ma the Oldowan diffused from a single geographic origin, with technological variants emerging in response to local ecological conditions of raw material availability. When the relative simplicity of Oldowan stone knapping is combined with the known sporadic distribution of Oldowan occurrences, especially pre-2.0 Ma, the question arises whether the long time span of the Oldowan encompasses recurrent discoveries of the basic principles of percussion flaking or diffusion of a material culture technology through transmission from a single origin.

Some explanations for lack of culture growth and change in human populations reside in the interaction of demographic and cognitive factors (Enquist et al., 2008). Once cultural traits are established, some level of demographic stability is essential for maintaining them. That is because the traits are carried and transferred by individuals through some form of social learning, and the number of cultural traits within a population depends on the number of individuals from which others may learn. The diversity of cultural traits present in human populations is expected to increase with community size (e.g., Powell et al., 2009), and the modelled relationship between population size and cultural diversity is expected to hold if all members of the community participate equally in the transmission of cultural behaviours. This is not always the case. For example, among chimpanzees, females seem to be more important than males in transmitting cultural traits, including tool use (Lind and Lindenfors, 2010) because they have a stronger proclivity to engage in tool use, and because they spend more time with their offspring than do the males, thus creating more opportunities for vertical transmission of cultural traits or information (Boesch, 1991; Boesch and Boesch, 1990; Lonsdorf et al., 2004; van Schaik and Pradhan, 2003). Another mechanism that increases cultural diversity might be the mobility of individuals across groups who are able to introduce traits that are new to the recipient group. Again, among chimpanzees, females rather than males tend to transfer between communities (e.g., Gagneux et al., 1999; Goodall, 1986) and therefore have better opportunities for transmitting cultural traits across groups. Thus, cultural transmission within and between groups of chimpanzees does not depend on overall group size, but on the relative and absolute number of females within the group.

In principle, greater cultural diversity can create a greater potential for variation and culture growth. At the same time, individuals within a population who had established cultural traits may hold such growth at bay. As noted experimentally for captive chimpanzees (Hrubesch et al., 2009), individuals who became specialised in carrying out certain technological procedures could turn highly conservative and resist change regardless of its potential efficiency. Also, individuals who acquire information culturally may do so at the expense of individual creativity (Boyd and Richerson, 1985), which would then stifle the number of variants within a community and constrain cultural growth (Enquist et al., 2008).

In essence, our questions here are whether Oldowan groups were sufficiently stable demographically to maintain knowledge over long periods of time; whether the groups were structured enough for individuals to impose on their communities their personal dislike for novelties (which would keep the fidelity of existing techniques and procedures, but would suppress creativity); and whether between-group interactions occurred often enough for individuals to transfer technological knowledge from a donor group to a recipient group and to spread the Oldowan across East Africa from its origin in northeastern Ethiopia. (The problem is exacerbated if one allows that more than one hominin genus made stone tools, as this would make cultural transmission less likely.) This question, pertinent to any prehistoric period, seems more acute for the Oldowan, given the characteristics of the archaeological record, which indicate ephemeral and short-lived occurrences, and the time depth for which stasis is suggested by Stout et al. (2010).

To uphold a scenario of diffusion from a single source with variations that are nondirectional, it should be assumed that Oldowan groups were sufficiently stable so that stochastic mortality events did not eliminate the reservoir of technological (or other cultural) knowledge for a period of million years. It should also be assumed that mechanisms of cultural transmission – either similar to or different from those of extant chimpanzees – existed across a large geographic area.

The archaeological and fossil records (see Section 3) are moot with regard to such questions. Although certain aspects of primate anatomy are highly correlated with their social behaviour (e.g., the levels of sexual dimorphism), the fossil record of Oldowan toolmakes between 2.5 and 1.8 Ma is simply too scanty to tell us whether their social organisation was similar to that of extant chimpanzees, in which case similar mechanisms of transmission might be invoked.

Oldowan occurrences are few and widely spaced in time. Moreover, within the time of the Oldowan, each occurrence typically represents short-duration (on the scale of tens of years) occupations. Admittedly, these sites do not encompass all the activities of early hominins. Nor are these the only locations where social interactions occurred. Still, given that those localities are what defines the Oldowan as such, we need to consider the fact that this record does not depict unequivocal opportunities for stable cultural transmission systems over a one-million-year period.

On the other hand, chimpanzee material cultures and the archaeological record of recent periods, the two extremes of our range of permissible analogues for the Oldowan, caution against an unqualified model of continuous transmission of Oldowan lithic technological procedures. The material cultures of chimpanzees and orangutans occur among geographically discontinuous populations. While the capacity for some level of cultural behaviour may be inherent, the specific behaviours are more localised and are ascribed to the transmission and spread of specific inventions (Lycett et al., 2009, 2011; van Schaik et al., 2003). At the other extreme of the range of analogues, the record of modern humans in more recent archaeological periods shows that cultural innovations, for example, complex technologies such as heat treatment of stone or firing clay into pottery, each of which requires multiple steps and procedures, were sometimes invented, lost and independently reinvented in different geographic areas (e.g., Hovers and Belfer-Cohen, 2006; Jacobs et al., 2008; Shennan, 2001).

One has to allow for the possibility that such might have been the case with Oldowan stone toolmaking. The technological principles involved in lithic reduction in general, and the procedures of Oldowan lithic reduction in particular, are constrained by the natural properties of the raw material and the physics of conchoidal fracture mechanics. Hence, the range of solutions and of technological procedures available to the knapper were much smaller. The first Oldowan stone toolmaking involved some breach of a cognitive "glass ceiling", leading first to the invention of lithic knapping and making possible retention of the procedures. This cognitive capacity would enable hominins in the later Oldowan to regain the same or similar solutions. Rather than exploring Klondike spaces, hominins in the later part of the Oldowan would search for solutions within homing spaces. If reinventions occurred throughout the Oldowan, then philosophically and psychologically they would not have been acts of exceptional creativity (as defined by Boden, 1998; see above). Rather, they may be viewed as acts of mundane creativity.

Aspects of Oldowan technological organisation other than lithic technology show more pronounced changes from the earlier to later Oldowan. It has been argued (Potts, 1988, 1991) that the spatial clusters, which are the archaeological smoking gun of the Oldowan, could have emerged if early hominins transported spatially disjunctive lithic and subsistence resources and cached them in selected locations within their foraging range. This behaviour would be ecologically adaptive as it would render any movable resource within this range accessible for processing behaviours. This scenario was based on the earliest sites known at the time, dating 1.9 Ma or slightly later, to explain the emergence of the Oldowan (Potts, 1991). It does not withstand more recent data on older sites. The close proximity of pre-2.0-Ma sites to their raw material sources and the transport of artefacts and cores into these localities and away from them (Delagnes and Roche, 2005; Goldman-Neuman and Hovers, 2012; Hovers et al., 2008; Semaw et al., 2003; Toth et al., 2006) do not support the caching hypothesis for these localities.

This may have changed in the later Oldowan. Both the complexity of raw material selectivity and transport distances increased after ~2.0 Ma, with convincing evidence for intentional clustering of different types of lithic and subsistence resources (Domínguez-Rodrigo et al., 2007; de la Torre and Mora, 2005; Goldman-Neuman and Hovers, 2012; and references therein). This change in the spatial organisation of resource on the landscape appears to be a behavioural innovation of the later Oldowan (Potts, 1991). This shift seems concurrent with evidence for increased exploitation of faunal foods. Tooth-wear analyses support the view that diet breadth of early Oldowan toolmakers may not have differed from that of australopiths who did not use tools habitually. In contrast, the wear on teeth of East African *H. erectus* suggests that members of this taxon, known from ca. 1.8 Ma and one of the candidates for stone toolmaking in the later Oldowan, expanded their dietary breadth compared to other, earlier members of *Homo* (Ungar et al., in press).

Based on tooth-wear analyses, tools may have been used mainly to cut plant tissue in the context of plant exploitation (Ungar et al., in press) as well as in making tools. If this inference is accepted, the evidence for more habitual tool use for processing animal tissue, encountered in the record of the later Oldowan, is a late add-on to the Oldowan technological and behavioural repertoire. It would then attest to an economically useful innovation of the later Oldowan, but not to a radical or surprising invention. This inferred change would constitute an act of mundane, rather than exceptional, creativity.

Combined, these separate lines of evidence imply that in the later Oldowan several activities that had initially been unrelated to one another were purposefully tethered to single localities, sometimes at considerable transport costs. In the parlance of creativity research, each of the behaviours discussed here is an expansion of a previously known behaviour more than it is a radically novel combination of such behaviour. At best, these behaviours represent mundane creativity.

5.5. CONCLUDING REMARKS

The study of cognitive abilities of early hominins has always been an intriguing yet tricky domain of palaeoanthropological research, plagued by epistemological and analytical difficulties. The evolutionary distance between the cognition of modern humans and that of early stone toolmakers renders such discussions more complicated where the Oldowan is considered. In particular, creativity has been tough to tackle, because it is a cognitive capacity that is very strongly associated with modern humans (e.g., Coolidge and Wynn, 2009; Carruthers, 2002; Mithen, 1996; Noble and Davidson, 1996). Accordingly, it has not been high on the list of causes commonly implicated in the emergence of the Oldowan or its spread. Still, creativity should be viewed as one of the adaptive strategies of hominids in general, and hominins in particular.

The Oldowan archaeological record exhibits several patterns that are associated for the first time with hominins. A closer examination of this record shows that the majority of these are the continuation or extension of behavioural patterns that might have been shared by early hominins and nonhuman primates. Oldowan stone toolmaking is the only major disjunction from the toolmaking and/or tool-using behaviours of nonhuman primates and pre-Oldowan hominins (see also Lake, 1998). This first occurrence marks a creative act in the sense that it provided an unexpected and interesting solution to a problem faced by hominins. Considering the paucity of unambiguous evidence for meat consumption during the first 600,000 years of stone toolmaking, the makers and users of early Oldowan tools might have been searching for strategies to reduce risk (e.g., finding a way to enhance the supply of plant food staples) rather than for ways to intensify economic gains by obtaining more meat.

The first occurrence of the Oldowan – regardless of an "age tag" that future research may put on it – might correspond with a window of opportunity where cognitive changes in the organisation and capabilities of working memory enabled unorthodox use of known elements in the hominins' technological repertoire in unexpected ways. In tandem with changes in motor abilities and skill, this would have enabled the execution of early stone knapping as we know it.

The Oldowan time span has been perceived as a long period of technological stasis (Semaw, 2000). Certainly this is the case when compared to later periods in which changes are more recognisable and rates of change were faster. Whether the prevalence of the basic Oldowan technological procedures was the result of continuous cultural transmission or of reinventing/relearning is a question that will perhaps remain unanswerable. Still, the apparent homogeneity of the Oldowan stone-toolmaking technologies masks several changes in other aspects. Changes in land-use patterns and in dietary spectra seem to be late innovations within the Oldowan, and are not necessarily correlated with, or dependent on, the conservative procedures of stone-tool making.

Figuring out the mosaic of inventions and reinventions during the Oldowan and their demographic, social and cognitive implications, are some of the challenges of future research. Recognising that they exist is the first step in this research. The emergence of the Oldowan represents a momentous threshold in hominin evolution because it involved exceptional creativity, essentially different from cognitive patterns shared with other hominids. Innovations inferred from the Oldowan record after this event are attributed on the most part to mundane creativity, "…the workhorse that accounts for the bulk of human accomplishments…." (Barsalou and Prinz, 2000: 267). It is such innovations that eventually started far-reaching processes of technological and social organisation of hominin groups during and after the Oldowan.

ACKNOWLEDGEMENTS

I thank Scott Elias for inviting me to contribute to this volume and for his patience throughout the process. Anna Belfer-Cohen, Katharine McDonald and John Speth read and commented on earlier rough drafts of this paper. I thank them for their helpful suggestions, critiques and insights.

REFERENCES

Alemseged, Z., Bobe, R., 2009. Diet in early hominin species: a paleoenvironmental perspective. In: Hublin, J.-J., Richards, M.P. (Eds.), The Evolution of Hominid Diets: Integrating Approaches to the Study of Paleolithic Subsistence. Springer, Dordrecht, pp. 181–188.

Ambrose, S.H., 2001. Paleolithic technology and human evolution. Science 291, 1748–1753.

Andersson, C., 2011. Paleolithic punctuations and equilibria: did retention rather than innovation limit technological evolution? PaleoAnthropology 2011, 243-259. http://dx.doi.org/10.4207/PA.2011.ART55

Ashley, G.M., Barboni, D., Dominguez-Rodrigo, M., Bunn, H.T., Mabulla, A.Z.P., Diez-Martin, F., Barba, R., Baquedano, E., 2010. A spring and wooded habitat at FLK Zinj and their relevance to origins of human behavior. Quat. Res. 74, 304–314.

Backwell, L., d'Errico, F., 2003. Additional evidence on the early hominid bone tool from swartkrans with reference to spatial distribution of lithic and organic artefacts. S. Afr. J. Sci. 99, 259–267.

Bailey, G., 2007. Time perspectives, palimpsests and the archaeology of time. J. Anthropol. Archaeol. 26, 198–223.

Bania, A., Harris, S., Kinsley, H., Boysen, S., 2009. Constructive and deconstructive tool modification by chimpanzees *(Pan troglodytes)*. Anim. Cogn. 12, 85–95.

Barsalou, L.W., Prinz, J.J., 2000. Mundane creativity in perceptual symbol systems. In: Ward, T.B., Smith, S.M., Vaid, J. (Eds.), Creative Thought: An Investigation of Conceptual Structures and Processes. American Psychological Association, Washington, D.C, pp. 267–307.

Basalla, G., 1988. The Evolution of Technology. Cambridge University Press, Cambridge.

Belfer-Cohen, A., Hovers, E., 2010. Modernity, enhanced working memory, and the middle to upper Paleolithic record in the Levant. Current Anthropology 51, S167–S175.

Binford, L.R., 1981. Bones: Ancient Men and Modern Myths. Academic Press, New York.

Blumenschine, R.J., 1987. Characteristics of an early hominid scavenging niche. Current Anthropology 28, 383–407.

Blumenschine, R.J., 1991. Breakfast at Olorgesailie: the natural history approach to early Stone Age archaeology. Reivew of The Archaeology of Human Origins: Papers by Glynn Isaac Edited by Barbara Isaac. J. Hum. Evol. 21, 307–327. Cambridge University Press, Cambridge.

Blumenschine, R.J., 1995. Percussion marks, tooth marks, and experimental determinations of the timing of hominid and carnivore access to long bones at FLK Zinjanthropus, Olduvai Gorge, Tanzania. J. Hum. Evol. 29, 21–51.

Blumenschine, R.J., Cavallo, A., Capaldo, S.D., 1994. Competition for carcasses and early hominid behavioral ecology: a case study and conceptual framework. J. Hum. Evol. 27, 127–213.

Blumenschine, R.J., Prassack, K.A., Kreger, C.D., Pante, M.C., 2007. Carnivore tooth-marks, microbial bioerosion, and the invalidation of test of Oldowan hominin scavenging behavior. J. Hum. Evol. 53, 420–426.

Blumenschine, R.J., Masao, F.T., Tactikos, J.C., Ebert, J.I., 2008. Effects of distance from stone source on landscape-scale variation in Oldowan artifact assemblages in the Paleo-Olduvai Basin, Tanzania. J. Archaeol. Sci. 35, 76–89.

Boden, M.A., 1998. What is creativity? In: Mithen, S. (Ed.), Creativity in Human Evolution and Prehistory. Routeldge, London and New York, pp. 22–60.

Boesch, C., 1991. Teaching among wild chimpanzees. Anim. Behav. 41, 530–532.

Boesch, C., Boesch, H., 1989. Hunting behavior of wild chimpanzees in the Tai National Park. Am. J. Phys. Anthropol. 78, 547–573.

Boesch, C., Boesch, H., 1990. Tool use and tool making in wild chimpanzees. Folia. Primatol. 54, 86–99.

Boesch, C., Head, J., Robbins, M.M., 2009. Complex tool sets for honey extraction among chimpanzees in Loango National Park, Gabon. J. Hum. Evol. 56, 560–569.

Boyd, R., Richerson, P.J., 1985. Culture and the Evolutionary Process. University of Chicago Press, Chicago.

Brantingham, P.J., 1998. Hominid-carnivore coevolution and invasion of the predatory guild. J. Anthropol. Archaeol. 17, 327–353.

Braun, D.R., Hovers, E., 2009. Current issues in Oldowan research. In: Hovers, E., Braun, D.R. (Eds.), Interdisciplinary Approaches to the Oldowan. Springer, New York, pp. 1–14.

Braun, D.R., Pobiner, B.L., Thompson, J.C., 2008. An experimental investigation of cut mark production and stone tool attrition. J. Archaeol. Sci. 35, 1216–1223.

Braun, D.R., Plummer, T., Ferraro, J.V., Ditchfield, P., Bishop, L.C., 2009. Raw material quality and Oldowan hominin toolstone preferences: evidence from Kanjera South, Kenya. J. Archaeol. Sci. 36, 1605–1614.

Bunn, H.T., 1986. Patterns of skeletal representation and hominid subsistence activities at Olduvai |Gorge, Tanzania, and Koobi Fora, Kenya. J. Hum. Evol. 15, 673–690.

Bunn, H.T., 2007. Meat made us human. In: Ungar, P.S. (Ed.), Evolution of the Human Diet. The Known, the Unknown and the Unknowable. Oxford University Press, Oxford, pp. 191–211.

Bunn, H.T., Mabulla, A.Z.P., Domínguez-Rodrigo, M., Ashley, G.M., Barba, R., Diez-Martìn, F., Remer, K., Yravedra, J., Baquedano, E., 2010. Was FLK North levels 1-2 a classic "living floor" of Oldowan hominins or a taphonomically complex palimpsest dominated by large carnivore feeding behavior? Quat. Res. 74, 355–362.

Campbell, D.T., 1960. Blind variation and selective retention in creative thought as in other knowledge processes. Psychol. Rev. 67, 380–400.

Carruthers, P., 2002. Human creativity: its cognitive basis, its evolution, and its connections with childhood pretence. Br. J. Philos. Sci. 53, 225–249.

Carvalho, S., Cunha, E., Sousa, C.u., Matsuzawa, T., 2008. Chaines operatoires and resource-exploitation strategies in chimpanzee (*Pan troglodytes*) nut cracking. J. Hum. Evol. 55, 148–163.

Cerling, T.E., Mbua, E., Kirera, F.M., Manthi, F.K., Grine, F.E., Leakey, M.G., Sponheimer, M., Uno, K.T., 2011. Diet of Paranthropus boisei in the early Pleistocene of East Africa. Proc. Natl. Acad. Sci. U S A 108, 9337–9341.

Chavaillon, J., 1976. Evidence for the technical practices of early Pleistocene hominids, Shungura formation, Lower Omo Valley, Ethiopia. In: Coppens, Y., Howell, F.C., Isaac, G.L., Leakey, R.E.F. (Eds.), Earliest Man and Environments in the Lake Rudolf Basin. Chicago University Press, Chicago, pp. 565–573.

Chavaillon, J., Piperno, M., 2004. Studies on the Early Paleolithic Site of Melka Kunture, Ethiopia. Instituto Italiano di Preistoria e Protostoria, Florence.

Clark, J.D., 1958. The natural fracture of pebbles from the Batoka Gorge, Northern Rhodesia, and its bearing on the Kafuan industries of Africa. Proc. Prehistoric Soc. 4, 64–77.

Coolidge, F.L., Wynn, T., 2009. The Rise of *Homo sapiens*. The Evolution of Modern Thinking. Wiley-Blackwell.

Copeland, S.R., 2009. Potential hominin plant foods in northern Tanzania: semi-arid savannas versus savanna chimpanzee sites. J. Hum. Evol. 57, 365–378.

d'Errico, F., Backwell, L.R., Berger, L.R., 2001. Bone tool use in termite foraging by early hominids and its impact on our understanding of early hominid behaviour. S. Afr. J. Sci. 97, 71–75.

Davidson, I., McGrew, W.C., 2005. Stone tools and the uniqueness of human culture. J. R. Anthropol. Inst. 11, 793–817.

de Beaune, S.A., 2004. The invention of technology: prehistory and cognition. Current Anthropology 45, 139–162.

de Heinzlein, J., Clark, D.J., White, T., Hart, W., Renne, P.R., WoldeGabriel, G., Beyene, Y., Vrba, E., 1999. Environment and behavior of 2.5 million-year-old Bouri hominids. Science 284, 625–629.

de la Torre, I., Mora, R., 2005. Unmodified lithic material at Olduvai Bed I: manuports or ecofacts? J. Archaeol. Meth. Theor. 32, 273–285.

de la Torre, I., Mora Torcal, R., 2005. Technological Strategies in the Lower Pleistocene at Olduvai Beds I & II. Université de Liège. ERAUL 112, Liège.

Delagnes, A., Roche, H., 2005. Late Pliocene hominid knapping skills: the case of Lokalalei 2C, West Turkana, Kenya. J. Hum. Evol. 48, 435–472.

Delagnes, A., Boisserie, J.-R., Beyene, Y., Chuniaud, K., Guillemot, C., Schuster, M., 2011. Archaeological investigations in the lower Omo Valley (Shungura Formation, Ethiopia): new data and perspectives. J. Hum. Evol. 61, 215–222.

Dennett, D.C., 1997. The intentional stance in theory and practice. In: Byrne, R.W., Whiten, A. (Eds.), Machiavellian Intelligence. Social Expertise and the Evolution of Intellect in Monkeys, Apes, and Humans. Clarendon Press, Oxford, pp. 180–202.

Dirks, P.H.G.M., Kibii, J.M., Kuhn, B.F., Steininger, C., Churchill, S.E., Kramers, J.D., Pickering, R., Farber, D.L., Meriaux, A.-S., Herries, A.I.R., King, G.C.P., Berger, L.R., 2010. Geological setting and age of *Australopithecus sediba* from southern Africa. Science 328, 205–208.

Domínguez-Rodrigo, M., 2002. Hunting and scavenging by early humans: the state of the debate. J. World Prehistory 16, 1–54.

Domínguez-Rodrigo, M., 2009. Are all Oldowan sites palimpsests? If so, what can they tell us about hominin carnivory? In: Hovers, E., Braun, D.R. (Eds.), Interdisciplinary Approaches to the Oldowan. Springer, New York, pp. 129–148.

Domínguez-Rodrigo, M., Barba, R., 2006. New estimates of tooth mark and percussion mark frequencies at the FLK Zinj site: the carnivore-hominid-carnivore hypothesis falsified. J. Hum. Evol. 50, 170–194.

Domínguez-Rodrigo, M., Martínez-Navarro, B., 2012. Taphonomic analysis of the early Pleistocene (2.4 Ma) faunal assemblage from A.L. 894 (Hadar, Ethiopia). J. Hum. Evol., 62, 315-327.

Domínguez-Rodrigo, M., Pickering, T.R., Semaw, S., Rogers, M.J., 2005. Cutmarked bones from Pliocene archaeological sites at Gona, Afar, Ethiopia: implications for the function of the world's oldest stone tools. J. Hum. Evol. 48, 109–121.

Domínguez-Rodrigo, M., Barba, R., Egeland, C.P., 2007. Deconstructing Olduvai. A Taphonomic Study of the Bed I Sites. Springer, Dordrecht, The Netherlands.

Domínguez-Rodrigo, M., Bunn, H.T., Mabulla, A.Z.P., Ashley, G.M., Diez-Martin, F., Barboni, D., Prendergast, M.E., Yravedra, J., Barba, R., Sanchez, A., Baquedano, E., Pickering, T.R., 2010. New excavations at the FLK Zinjanthropus site and its surrounding landscape and their behavioral implications. Quat. Res. 74, 315–332.

Egeland, C.P., Pickering, T.R., Domínguez-Rodrigo, M., Brain, C.K., 2004. Disentangling early stone age palimpsests: determining the functional independence of hominid- and carnivore-derived portions of archaeofaunas. J. Hum. Evol. 47, 343–357.

Enquist, M., Ghirlanda, S., Jarrick, A., Wachtmeister, C.A., 2008. Why does human culture increase exponentially? Theor. Popul. Biol. 74, 46–55.

Feibel, C.S., Brown, F., McDougal, I., 1989. Stratigraphic context of fossil hominids from the Omo group deposits: northern Turkana Basin, Kenya and Ethiopia. Am. J. Phys. Anthropol. 78, 595–622.

Ferring, R., Oms, O., Agustí, J., Berna, F., Nioradze, M., Shelia, T., Tappen, M., Vekua, A., Zhvania, D., Lordkipanidze, D., 2011. Earliest human occupations at Dmanisi (Georgian Caucasus) dated to 1.85-1.78 Ma. Proc. Natl. Acad. Sci. U S A 108, 10432–10436.

Fragaszy, D., Isar, P., Visalberghi, E., Ottoni, E.B., De Oliveira, M.G., 2004. Capuchin monkeys (Cebus libidinosus) use anvils and stone pounding tools. Am. J. Primatol. 64, 359–366.

Gagneux, P., Boesch, C., Woodruff, D.S., 1999. Female reproductive strategies, paternity and community structure in wild West African chimpanzees. Anim. Behav. 57, 19–32.

Gamble, C., 2007. Origins and Revolutions. Human Identity in Earliest Prehistory. Cambridge University Press, Cambridge.

Gardner, H., 1983. Frames of Mind: The Theory of Multiple Intelligence. Basic Books, New York.

Gibbons, A., 2009. Of tools and tubers. Science 324, 588–589.

Gold, R., 2007. Creativity, Innovation, and Making Stuff. MIT Press, Cambridge MA.

Goldman-Neuman, T., Hovers, E., 2009. Methodological considerations in the study of Oldowan raw material selectivity: insights from A. L. 894 (Hadar, Ethiopia). In: Hovers, E., Braun, D.R. (Eds.), Interdisciplinary Approaches to the Oldowan. Springer, New York, pp. 71–84.

Goldman-Neuman, T., Hovers, E., 2012. Raw material selectivity in the Late Pliocene Oldowan sites of the Makaamitalu Basin, Hadar. J. Hum. Evol. 62, 353-366

Goodall, J., 1986. The Chimpanzees of Gombe. Patterns of Behavior. Belknap, Cambridge, MA and London.

Gould, S.J., 1986. Evolution and the triumph of homology, or why history matters. Am. Sci. 74, 60–69.

Harmand, S., 2009. Variability in raw material selection at the late pliocene sites of Lokalalei, West Turkana, Kenya. In: Hovers, E., Braun, D.R. (Eds.), Interdisciplinary Approaches to the Oldowan. Springer, Dordrecht, The Netherlands, pp. 85–98.

Harris, J.W.K., Capaldo, S.D., 1993. The earliest stone tools: their implications for an understanding of the activities and behaviour of late Pliocene hominids. In: Berthelet, A., Chavaillon, J. (Eds.), The Use of Tools by Human and Nonhuman Primates. Clarendon Press, Oxford, pp. 197–224.

Harris, J.W.K., Williamson, P.G., Verniers, J., Tappen, M.J., Stewart, K., Helgren, D., de Heinzelin, J., Boaz, N.T., Bellomo, R.V., 1987. Late Pliocene hominid occupation in Central Africa: the setting, context and character of the Senga 5A site, Zaire. J. Hum. Evol. 16, 701–728.

Hernandez-Aguilar, R.A., 2009. Chimpanzee nest distribution and site reuse in a dry habitat: implications for early hominin ranging. J. Hum. Evol. 57, 350–364.

Hernandez-Aguilar, R.A., Moore, J., Pickering, T.R., 2007. Savanna chimpanzees use tools to harvest the underground storage organs of plants. Proc. Natl. Acad. Sci. U S A 104, 19210–19213.

Hovers, E., 2009a. The Lithic Assemblages of Qafzeh Cave. Oxford University Press, New York.

Hovers, E., 2009b. Learning from mistakes: flaking accidents and knapping skills in the assemblage of A. L. 894 (Hadar, Ethiopia). In: Schick, K., Toth, N. (Eds.), The Cutting Edge: New Approaches to the Archaeology of Human Origins. Stone Age Institute, Gosport, pp. 137–150.

Hovers, E., Belfer-Cohen, A., 2006. "Now you see it, now you don't" – modern human behavior in the middle Paleolithic. In: Hovers, E., Kuhn, S.L. (Eds.), Transitions Before the Transition: Evolution and Stability in the Middle Paleolithic and Middle Stone Age. Springer, New York, pp. 295–304.

Hovers, E., Vandermeersch, B., Bar-Yosef, O., 1997. A middle Palaeolithic engraved artefact from Qafzeh Cave, Israel. Rock Art Res. 14, 79–87.

Hovers, E., Campisano, C.J., Feibel, C.S., 2008. Land use by Late Pliocene hominins in the Makaamitalu Basin, Hadar, Ethiopia. Paleoanthropology 2008, A14.

Hovers, E., Davidzon, A. in preparation. Late pliocene Oldowan lithic technology from the Makaamitalu Basin localities, Hadar, Ethiopia.

Howell, F.C., Haesaerts, P., de Hienzlin, P., 1987. Depositional environments, archaeological occurrences and hominids from members E and F of the Shungura formation (Omo Basin, Ethiopia). J. Hum. Evol. 16, 665–700.

Hrubesch, C., Preuschoft, S., van Schaik, C., 2009. Skill mastery inhibits adoption of observed alternative solutions among chimpanzees (Pan troglodytes). Anim. Cogn. 12, 209–216.

Isaac, G.L., 1971. The diet of early man: aspects of archaeological evidence from lower and middle Pleistocene sites in Africa. World Archaeol. 2, 278–299.

Isaac, G.L., 1976. Plio-Pleistocene artifact assemblages from East Rudolf, Kenya. In: Coppens, Y., Howell, F.C., Isaac, G.L., Leakey, R. (Eds.), Earliest Man and Environments in the Lake Rudolf Basin: Stratigraphy, Paleoecology, and Evolution. Univeristy of Chicago Press, Chicago, pp. 552–564.

Isaac, G.L., 1978. The food sharing behavior of proto-human hominids. Sci. Am. 238, 90–108.

Isaac, G.L., 1981. Stone Age visiting cards: approaches to the study of early land-use patterns. In: Hodder, I., Isaac, G.L. (Eds.), Patterns of the Past: Studies in Honor of David Clarke. Cambridge University Press, Cambridge, pp. 131–155.

Isaac, G.L., 1983. Bones in contention: competing explanations for the juxtaposition of Early Pleistocene artefacts and faunal remains. In:

Clutton-Brock, J., Grigson, C. (Eds.), Animals in Archaeology: Hunters and their Prey. BAR International Series, vol. 163. British Archaeological Reports, Oxford, pp. 3–19.

Isaac, G.L., 1984. The archaeology of human origins: studies of the Lower Pleistocene of East Africa. In: Wendorf, F., Close, A. (Eds.), Advances in Old World Archaeology, vol. 3. Academic Press, New York, pp. 1–87.

Isaac, G.L., Crader, D.C., 1981. To what extent were early hominids carnivorous? An archaeological perspective. In: Harding, R.S.O., Teleki, G. (Eds.), Omnivorous Primates: Gathering and Hunting in Human Evolution. Columbia press, New York, pp. 37–103.

Isaac, G.L., Harris, J.W.K., 1978. The Archaeological context of the hominid fossils. In: Leakey, M.G., Leakey, F.E.F. (Eds.), Koobi Fora Research Project Series, vol. 1. Clarendon Press, oxford, pp. 64–85.

Isaac, G.L., Harris, J.W.K., 1980. A method for determining the characteristics of artifacts between sites in the upper member of the Koobi Fora Formation, East Turkana, In: Leakey, M.G., Leakey, R.E.F. (Eds), Proceedings of the 8th Panafrican Congress of Prehistory and Quaternary Studies. Nairobi, pp. 19–22.

Isaac, G.L., Isaac, B., 1997. Koobi Fora Research Project Vol. 5: Plio-Pleistocene Archaeology. Clarendon Press, Oxford.

Jacobs, Z., Roberts, R.G., Galbraith, RF., Deacon, HJ., Grun, R., Mackay, A., Mitchell, P., Vogelsang, R., Wadley, L., 2008. Ages for the Middle Stone Age of southern Africa: implications for human behavior and dispersal. Science 322, 733–735.

Jones, P.R., 1994. Results of experimental work in relation to the stone industries of Olduvai Gorge. In: Leakey, M.D., Roe, D.A. (Eds.), Olduvai Gorge Vol. 5: Excavations in Beds III and IV, and the Masek Beds 1968–1971. Cambridge University Press, Cambridge, pp. 254–298.

Keeley, L., Toth, N., 1981. Microwear polishes on early stone tools from Koobi Fora, Kenya. Nature 293, 464–465.

Kimbel, W.H., Walter, R.C., Johanson, D.C., Reed, K.E., Aronson, J.L., Assefa, Z., Marean, C.W., Eck, G.G., Bobe, R., Hovers, E., Rak, Y., Vondra, C., Yemane, T., York, D., Chen, Y., Evensen, N.M., Smith, P.E., 1996. Late pliocene *homo* and Oldowan stone tools from the Hadar formation (Kada Hadar Member), Ethiopia. J. Hum. Evol. 31, 549–561.

Kitahara-Frisch, J., 1993. The origin of secondary tools. In: Berthelet, A., Chavaillon, J. (Eds.), The Use of Tools by Human and Non-Human Primates. Clarendon Press, Oxford, pp. 239–248.

Kivell, T.L., Kibii, J.M., Churchill, S.E., Schmid, P., Berger, L.R., 2011. Australopithecus sediba hand demonstrates mosaic evolution of locomotor and manipulative abilities. Science 333, 1411–1417.

Kortland, A., 1986. The use of stone tools by wild-living chimpanzees and earliest hominids. Journal of Human Evolution 15, 77–132.

Kuhn, S.L., Hovers, E., 2006. General introduction. In: Hovers, E., Kuhn, S.L. (Eds.), Transitions before the Transition: Evolution and Stability in the Middle Paleolithic and Middle Stone Age. Springer, New York, pp. 1–11.

Kuhn, S.L., Stiner, M.C., 1998. Middle Paleolithic 'creativity': reflections on an oxymoron? In: Mithen, S. (Ed.), Creativity in Human Evolution and Prehistory. Routledge, London and New York, pp. 143–164.

Lake, M., 1998. 'Homo': the creative genus? In: Mithen, S. (Ed.), Creativity in Human Evolution and Prehistory. Routledge, London, pp. 125–142.

Laland, K.N., Odling-Smee, J., Feldman, M.W., 2000. Niche construction, biological evolution, and cultural change. Behav. Brain. Sci. 23, 131–175.

Laland, K., O'Brien, M., 2010. Niche construction theory and archaeology. J. Archaeol. Meth. Theor. 17, 303–322.

Leakey, M.D., 1971. Olduvai Gorge Vol. 3, Excavations in Beds I and II, 1960–1963. Cambridge University Press, Cambridge.

Lee-Thorp, J.A., Sponheimer, M., Passey, B.H., de Ruiter, D.J., Cerling, T.E., 2010. Stable isotopes in fossil hominin tooth enamel suggest a fundamental dietary shift in the Pliocene. Philos. Trans. R. Soc. Lond. B Biol. Sci. 365, 3389–3396.

Levin, N.E., Quade, J., Simpson, S.W., Semaw, S., Rogers, M., 2004. Isotopic evidence for Plio-Pleistocene environmental change at Gona, Ethiopia. Earth. Planet. Sci. Lett. 219, 93–110.

Lind, J., Lindenfors, P., 2010. The number of cultural traits is correlated with female group size but not with male group size in chimpanzee communities. PLoS ONE 5, e9241. http://dx.doi.org/10.1371/journal.pone.0009241.

Lonsdorf, E.V., Pusey, A.E., Eberly, L., 2004. Sex differences in learning in chimpanzees. Nature 428, 715–716.

Lycett, S.J., Collard, M., McGrew, W.C., 2009. Cladistic analyses of behavioural variation in wild *Pan troglodytes*: exploring the chimpanzee culture hypothesis. J. Hum. Evol. 57, 337–349.

Lycett, S.J., Collard, M., McGrew, W.C., 2011. Correlations between genetic and behavioural dissimilarities in wild chimpanzees (*Pan troglodytes*) do not undermine the case for culture. Proc. R. Soc. Lond. B Biol. Sci. 278, 2091–2093.

Malinsky-Buller, A., Hovers, E., Marder, O., 2011. Making time: 'living floors', 'palimpsests' and site formation processes – a perspective from the open-air lower Paleolithic site of revadim quarry, Israel. J. Anthropol. Archaeol. 30, 89–101.

Marzke, M.W., 1997. Precision grips, hand morphology, and tools. American Journal of Physical Anthropology 102, 91–110.

McGrew, W.C., 1992. Chimpanzee Material Culture: Implications for Human Evolution. Cambridge University Press, Cambridge.

McGrew, W.C., 1993. The intelligent use of tools: twenty proposals. In: Gibson, K.R., Ingold, T. (Eds.), Tools, Language and Cognition in Human Evolution. Cambridge Universtiy Press, Cambrige, pp. 151–170.

McPherron, S.P., Alemseged, Z., Marean, C.W., Wynn, J.G., Reed, D., Geraads, D., Bobe, R., Bearat, H.A., 2010. Evidence for stone-tool-assisted consumption of animal tissues before 3.39 million years ago at Dikika, Ethiopia. Nature 466, 857–860.

Mercader, J., Panger, M., Boesch, C., 2002. Excavation of a chimpanzee stone tool site in the African rain forest. Science 296, 1452–1455.

Mercader, J., Barton, H., Gillespie, J., Harris, J., Kuhn, S., Tyler, R., Boesch, C., 2007. 4,300-year-old chimpanzee sites and the origins of percussive stone technology. Proc. Natl. Acad. Sci. U S A 104, 3043–3048.

Mithen, S., 1996. The Prehistory of the Mind: A Search for the Origins of Art, Religion and Science. Thames and Hudson.

Mithen, S., 1998a. Introduction. In: Mithen, S. (Ed.), Creativity in Human Evolution and Prehistory. Routeledge, London, pp. 1–15.

Mithen, S., 1998b. A creative explosion? theory of mind, language and the disembodied mind of the upper Paleolithic. In: Mithen, S. (Ed.), Creativity in Human Evolution and Prehisotry. Routeledge, London, pp. 165–191.

Moore, M.W., 2010. "Grammars of action" and stone flaking design space. In: Nowell, A., Davidson, I. (Eds.), Stone Tools and the Evolution of Human Cognition. University Press of Colorado, Boulder, Colorado, pp. 13–43.

Morgan, B.J., Abwe, E.E., 2006. Chimpanzees use stone hammers in Cameroon. Curr. Biol. 16, R632–R633.

Moura, A.C., Lee, P.C., 2004. Capuchin stone tool use in Caatinga dry forest. Science 306, 1909.

Mulcahy, N.J., Call, J., 2006. Apes save tools for future use. Science 312, 1038–1040.

Noble, W., Davidson, I., 1996. Human Evolution, Language and Mind. Cambridge University Press, Cambridge.

Osvath, M., 2009. Spontaneous planning for future stone throwing by a male chimpanzee. Curr. Biol. 19, R190–R191.

Osvath, M., 2010. Great ape foresight is looking great. Anim. Cogn. 13, 777–781.

Osvath, M., Gardenfors, P., 2005. Oldowan Culture and the Evolution of Anticipatory Cognition. Lund University Cognitive Science, Lund. http://wwwlucsluse/LUCS/.

Osvath, M., Osvath, H., 2008. Chimpanzee (*Pan troglodytes*) and orangutan (*Pongo abelii*) forethought: self-control and pre-experience in the face of future tool use. Anim. Cogn. 11, 661–674.

Panger, M.A., Brooks, A.S., Richmond, B.G., Wood, B., 2002. Older than the Oldowan? Rethinking the emergence of hominin tool use. Evol. Anthropol. Issues News Rev. 11, 235–245.

Perkins, D.N., 1992. The topography of invention. In: Weber, R.J., Perkins, D.N. (Eds.), Inventive Minds: Creativity in Technology. Oxford University Press, New York, pp. 238–250.

Perkins, D.N., 2000. Creativity's camel: the role of analogy in invention. In: Ward, T.B., Smith, S.M., Vaid, J. (Eds.), Creative Thought: An Investigation of Conceptual Structures and Processes. American Psychological Association, Washington, D.D, pp. 523–538.

Petraglia, M.D., Potts, R., 1994. Water flow and the formation of early Pleistocene artifact sites in Olduvai Gorge, Tanzania. J. Anthropol. Archaeol. 13, 228–254.

Pickering, R., Dirks, P.H.G.M., Jinnah, Z., de Ruiter, D.J., Churchill, S.E., Herries, A.I.R., Woodhead, J.D., Hellstrom, J.C., Berger, L.R., 2011. *Australopithecus sediba* at 1.977 Ma and implications for the origins of the genus *Homo*. Science 333, 1421–1423.

Plummer, T.W., 2004. Flaked stones and old bones: biological and cultural evolution at the dawn of technology. Yearb. Phys. Anthropol. 47, 118–164.

Plummer, T.W., Stanford, C.B., 2000. Analysis of a bone assemblage made by chimpanzees at gombe national park, Tanzania. J. Hum. Evol. 39, 345–365.

Plummer, T.W., Bishop, L.C., Ditchfield, P., Hicks, J., 1999. Research on late pliocene Oldowan sites at Kanjera South, Kenya. J. Hum. Evol. 36, 151–170.

Plummer, T.W., Ditchfield, P.W., Bishop, L.C., Kingston, J.D., Ferraro, J.V., Braun, D.R., Hertel, F., Potts, R., 2009. Oldest evidence of toolmaking hominins in a grassland-dominated ecosystem. PLoS ONE 4, e7199.

Potts, R., 1988. Early Hominid Activities at Olduvai. Aldine, New York.

Potts, R., 1991. Why the Oldowan? Plio-Pleistocene toolmaking and the transport of resources. J. Anthropol. Res. 47, 156–176.

Powell, A., Shennan, S., Thomas, M.G., 2009. Late Pleistocene demography and the appearance of modern human behavior. Science 324, 1298–1301.

Prat, S., Brugal, J.-P., Tiercelin, J.-J., Barrat, J.-A., Bohn, M., Delagnes, A., Harmand, S., Kimeu, K., Kibunjia, M., Texier, P.-J., Roche, H., 2005. First occurrence of early *Homo* in the nachukui formation (West Turkana, Kenya) at 2.3–2.4 Myr. J. Hum. Evol. 42, 230–240.

Pruetz, J.D., Bertolani, P., 2007. Savanna chimpanzees, *Pan troglodytes verus*, hunt with tools. Curr. Biol. 17, 412–417.

Ragir, S., Rosenberg, M., Tiemo, P., 2000. Gut morphology and the avoidance of carrion among chimpanzees, baboons, and early hominids. J. Anthropol. Res. 56, 477–512.

Read, D., 2008. Working memory: a cognitive limit to nonhuman primate recursive thinking prior to hominid evolution. Evol. Psychol. 6 (4), 676–714.

Read, D., van der Leeuw, S., 2008. Biology is only part of the story. Philos. Trans. R. Soc. Lond. B Biol. Sci. 363, 1959–1968.

Reed, D.N., Geraads, D., 2012. Evidence for a late Pliocene faunal transition based on a new rodent assemblage from Oldowan locality Hadar A.L. 894, afar region, Ethiopia. J. Hum. Evol, 62, 328-337.

Renfrew, C., 1978. The anatomy of innovation. In: Renfrew, C. (Ed.), Approaches to Social Archaeology. Harvard University Press, Cambridge, pp. 390–418.

Rightmire, G.P., Lordkipanidze, D., Vekua, A., 2006. Anatomical descriptions, comparative studies and evolutionary significance of the hominin skulls from dmanisi, Republic of Georgia. J. Hum. Evol. 50, 115–141.

Roche, H., Brugal, J.-P., Delagnes, A., Feibel, C., Harmand, S., Kibunjia, M., Prat, S., Texier, P.-J., 2003. Les sites archéologiques plio-plèistocènes de la formation de nachukui, Ouest-Turkana, Kenya: bilan synthétique 1997–2001. Comptes rendus Palevol 2, 663–673.

Rogers, M.J., Semaw, S., 2009. From nothing to something: the appearance and context of the earliest archaeological record. In: Camps, M., Chauhan, P.R. (Eds.), Sourcebook of Paleolithic Transitions. Methods, Theories, and Interpretations. Springer, New York, pp. 155–171.

Rogers, M.J., Harris, J.W.K., Feibel, C.S., 1994. Changing patterns of land use by Plio-Pleistocene hominids in the Lake Turkana Basin. J. Hum. Evol. 27, 139–158.

Rolian, C., Lieberman, D.E., Zermeno, J.P., 2011. Hand biomechanics during simulated stone tool use. Journal of Human Evolution 61, 26–41.

Sahnouni, M., 2006. The North African Early Stone Age and the sites at Ain anech, Algeria. In: Toth, N., Schick, K.D. (Eds.), The Oldowan: Case Studies into the Earliest Stone Age. Stone Age Institute Press, Bloomington, pp. 77–111.

Sahnouni, M., Schick, K.D., Toth, N., 1997. An experimental investigation into the nature of faceted limestone "spheroids" in the early Palaeolithic. J. Archaeol. Sci. 24, 701–713.

Sahnouni, M., Hadjouis, D., Van Der Made, J., Derradji, A., Canals, A., 2002. Further research at the Oldowan site of ain Hanech, North-east Algeria. J. Hum. Evol. 43, 925–937.

Sahnouni, M., Van der Made, J., Everett, M., 2011. Ecological background to Plio-Pleistocene hominin occupation in North Africa: the vertebrate faunas from ain boucherit, ain hanech and el-kherba, and paleosol stable-carbon-isotope studies from El-Kherba, Algeria. Quat. Sci. Rev. 30, 1303–1317.

Sakura, O., Matsuzawa, T., 1991. Flexibility of wild chipanzee nut-cracking behavior using using stone hammers and anvils: an experimental analysis. Ethology 87, 237–248.

Sanz, C.M., Morgan, D.B., 2007. Chimpanzee tool technology in the goualougo triangle, Republic of Congo. J. Hum. Evol. 52, 420–433.

Sanz, C., Morgan, D., 2009. Flexible and persistent tool-using strategies in honey-gathering by wild chimpanzees. Int. J. Primatol. 30, 411–427.

Sanz, C., Morgan, D., Gulick, S., 2004. New insights into chimpanzees, tools, and termites from the Congo Basin. Am. Nat. 164, 567–581.

Schick, K.D., 1987. Modeling the formation of Early Stone Age artifact concentrations. J. Hum. Evol. 16, 789–807.

Schick, K.D., Toth, N., Garufi, G., Savage-Rumbaugh, E.S., Raumbaugh, D., Sevick, R., 1999. Continuing investigations into the stone tool-making and tool-using capabilities of a Bonobo (Pan paniscus). J. Archaeol. Sci. 26, 821–832.

Semaw, S., 2000. The world's oldest stone artefacts from Gona, Ethiopia: their implications for understanding stone technology and patterns of human evolution between 2.6–1.5 million years ago. J. Archaeol. Sci. 27, 1197–1214.

Semaw, S., Rogers, M.J., Quade, J., Renne, P.R., Butler, R.F., Dominguez-Rodrigo, M., Stout, D., Hart, W.S., Pickering, T., Simpson, S.W., 2003. 2.6-Million-year-old stone tools and associated bones from OGS-6 and OGS-7, gona, afar, Ethiopia. J. Hum. Evol. 45, 169–177.

Sept, J.M., 1986. Plant foods and early hominids at site FxJj 50, Koobi Fora, Kenya. J. Hum. Evol. 15, 751–770.

Sept, J.M., 1992. Was there no place like home? a new perspective of early hominid archaeological sites from the mapping of chimpanzee nests. Curr. Anthropol. 33, 187–207.

Sept, J.M., 1994. Beyond bones: archaeological sites, early hominid subsistence, and the costs and benefits of exploiting wild plant foods in East African riverine landscapes. J. Hum. Evol. 27, 295–320.

Shennan, S., 2001. Demography and cultural innovation: a model and its implications for the emergence of modern human culture. Camb. Archaeol. J. 11, 5–16.

Simonton, D.K., 1999. Creativity as blind variation and selective retention: is the creative process Darwinian? Psychol. Inq. 10, 309–328.

Simonton, D.K., 2000. Human creativity, cultural evolution, and niche construction. Behav. Brain. Sci. 23, 159–160.

Speth, J.D., 2010. The Paleoanthropology and Archaeology of Big-game Hunting. Protein, Fat, or Politics? Springer, New York.

Spratt, D.A., 1982. The analysis of innovation processes. J. Archaeol. Sci. 9, 79–94.

Stanford, C.B., 1996. The hunting ecology of wild chimpanzees: implications for the behavioral ecology of Pliocene hominids. Am. Anthropol. 98, 96–113.

Stout, D., Quade, J., Semaw, S., Rogers, M.J., Levin, N.E., 2005. Raw material selectivity of the earliest stone toolmakers at gona, afar, Ethiopia. J. Hum. Evol. 48, 365–380.

Stout, D., Semaw, S., Rogers, M.J., Cauche, D., 2010. Technological variation in the earliest Oldowan from gona, afar, Ethiopia. J. Hum. Evol. 58, 474–491.

Sugiyama, Y., Koman, J., 1979. Tool-using and making- behavior in wild chipanzees in Bossou, Guinea. Primates 20, 513–524.

Tappen, M.J., Wrangham, R., 2000. Recognizing hominoid-modified bones: the taphonomy of *Colobus* bones partially digested by free-ranging chimpanzees in the kibale forest, Uganda. Am. J. Phys. Anthropol. 113, 217–234.

Texier, P.J., 1995. The Oldowan assemblage from NY 18 site at nyabusosi (Toro-Uganda). Comptes Rendus Acad. Sci. Paris 320, 647–653.

Tocheri, M.W., Orr, C.M., Jacofsky, M.C., Marzke, M.W., 2008. The evolutionary history of the hominin hand since the last common ancestor of *Pan* and *Homo*. J. Anat. 212, 544–562.

Tomasello, M., 1999. The Cultural Origins of Human Cognition. Cambridge University Press, Cambridge MA.

Toth, N., 1985. The Oldowan reassessed: a close look at early stone artifacts. J. Archaeol. Sci. 12, 101–120.

Toth, N., 1997. The artefact assemblages in the light of experimental studies. In: Isaac, G.L., Isaac, B. (Eds.), Koobi Fora Research Project. Vol. 5: Plio-Pleistocene Archaeology. Clarendon Press, Oxford, pp. 362–401.

Toth, N., Schick, K., 2009. The Oldowan: the tool making of early hominins and chimps compared. Annu. Rev. Anthropol. 38, 289–305.

Toth, N., Schick, K.D., Savage-Rumbaugh, E.S., Sevick, R.A., Rumbaugh, D.M., 1993. *Pan* the tool-maker: investigations into the stone tool-making and tool-using capabilities of a bonobo (*Pan paniscus*). J. Archaeol. Sci. 20, 81–91.

Toth, N., Schick, K.D., Semaw, S., 2006. A comparative study of the stone tool-making skills of Pan, Australopithecus and *Homo sapiens*. In: Toth, N., Schick, K.D. (Eds.), The Oldowan: Case Studies into the Earliest Stone Age. Stone Age Institute Press, Bloomington, pp. 155–222.

Ungar, P.S., Grine, F.E., Teaford, M.F., 2006. Diet in early *Homo*: a review of the evidence and a new model of adaptive versatility. Annu. Rev. Anthropol. 35, 209–228.

Ungar, P.S., Krueger, K.L., Blumenschine, R.J., Njau, J., Scott, R.S., in press. Dental microwear texture analysis of hominins recovered by the Olduvai landscape paleoanthropology project, 1995-2007. J. Hum. Evol. doi: 10.1016/j.jhevol.2011.04.006.

van der Leeuw, S.E., Torrence, R., 1989. What 's New? A Closer Look at the Process of Innovation. Unwin Hyman, London.

van der Merwe, N.J., Masao, F.T., Bamford, M.K., 2008. Isotopic evidence for contrasting diets of early hominins *Homo habilis* and Australopithecus boisei of Tanzania. S. Afr. J. Sci. 104, 153–155.

van Schaik, C.P., Knott, C.D., 2001. Geographic variation in tool use on *Neesia* fruits in orangutans. Am. J. Phys. Anthropol. 114, 331–342.

van Schaik, C.P., Pradhan, G.R., 2003. A model for tool-use traditions in primates: implications for the coevolution of culture and cognition. J. Hum. Evol. 44, 645–664.

van Schaik, C.P., Deaner, R.O., Merrill, M.Y., 1999. The conditions for tool use in primates: implications for the evolution of material culture. J. Hum. Evol. 36, 719–741.

van Schaik, C.P., Ancrenaz, M., Borgen, G., Galdikas, B., Knott, C.D., Singleton, I., Suzuki, A., Utami, S.S., Merrill, M., 2003. Orangutan cultures and the evolution of material culture. Science 299, 102–105.

Ward, T.B., Smith, S.M., Vaid, J., 2000. Conceptual structures and processes in creative thought. In: Ward, T.B., Smith, S.M., Vaid, J. (Eds.), Creative Thought: An Investigation of Conceptual Structures and Processes. American Psychological Association, Washington DC, pp. 1–26.

Westergaard, G.C., 1995. The stone tool technology of capuchin monkeys: possible implications for the evolution of symbolic communication in hominids. World Archaeol. 27, 1–9.

Wood, B., 1991. Koobi Fora Research Project Vol. 4: Hominid Cranial Remians. Clarendon Press, Oxford.

Wood, B., 1997. The oldest whodunnit in the world. Nature 385, 292–293.

Wrangham, R., Cheney, D., Seyfarth, R., Sarmiento, E., 2009. Shallow-water habitats as sources of fallback foods for hominins. Am. J. Phys. Anthropol. 140, 630–642.

Wynn, T., 1989. The Evolution of Spatial Competence. University of Illinois Press, Urbana and Chicago.

Ziman, J., 2000. Evolutionary models for technological change. In: Ziman, J. (Ed.), Technological Innovation as an Evolutionary Process. Cambridge University Press, Cambridge, pp. 3–12.

Chapter 6

Emergent Patterns of Creativity and Innovation in Early Technologies

Steven L. Kuhn

School of Anthropology, University of Arizona, Tucson, AZ 85721-0030 USA, skuhn@email.arizona.edu

6.1. INTRODUCTION

Human creativity can be recognised and studied at different scales. The most familiar use of the term refers to the individual act of producing something that is new. The artist painting feverishly in a drafty garret, the inventor experiencing an "ah ha" moment in a cluttered laboratory and the jazz musician on stage absorbed in a solo are the kinds of images that the word creativity prompts in most of us. However, we may also speak of creativity at very different human and temporal scales. Some communities, places and periods seem to be especially fertile sources of new inventions and artistic achievement: the European Renaissance and the Song Dynasty in China are some iconic examples. Others times and places, sometimes termed "dark ages", seem particularly sterile and monotonous by comparison.

If we recognise creativity at different scales, it follows that we must explain it in different ways. At the level of individual acts of genius, one is justified in looking to neurology, psychology and to the immediate social context. In trying to understand the creative fertility of particular periods or places, we need to think about more generalised social factors and, in the case of human evolution, about broad, population- or species- specific cognitive developments. These explanations do not compete – they refer instead to fundamentally different kinds of phenomena. The act of individual creation is the result of properties inherent in the individual interacting with local conditions. The phenomenon of a particularly creative period or place is an *emergent* property of the thoughts and actions of a great many different individuals; it is a consequence of the interactions more than individual characteristics.

The perceptual and analytical frames commonly adopted in Palaeoanthropology should encourage us to focus on some dimensions of the creative process while rendering others all but invisible. Palaeoanthropologists mark the passage of time in centuries or millennia, and they study human behaviour almost exclusively in the aggregate. In most situations, the results of individual acts of creation in the deep human past are essentially undetectable, swamped by the material remains of thousands of more mundane, repetitive acts. Moreover, even when they can be identified, products of singular genius are rare and unevenly distributed in time and space. Everyone would agree that the spectacular paintings in the Chauvet Cave (Clottes, 2003), the extraordinary ivory carvings and musical instruments from Aurignacian sites in Germany (Conard, 2009; Conard et al., 2009) or the engraved ostrich eggshell containers from Diepkloof rockshelter (Texier et al., 2010) bespeak a familiar form of creative expression that seems limited to humans among extant species. However, these are very rare things and they are confined largely, if not entirely, to that last phases of human prehistory, after the emergence of *Homo sapiens.*

Even if especially creative individuals and especially innovative objects are invisible in much of the compressed, time-transgressive record that Palaeoanthropologists study, we may still recognise the kinds of large-scale, emergent creativity that characterises particular places and periods. There are times and places in the past when a great many novel artefact forms or procedures appeared together or in rapid succession. The so-called "creative explosions" of the Eurasian Upper Palaeolithic or the African later MSA are prime examples, but there are other instances of localised "renaissances" or "golden ages". These seemingly fertile intervals are brought into sharp relief by comparison with times and places when very little that was new or different emerged.

The aim of this chapter is to explore alternative explanations for macroscopic, emergent patterns of creativity in human evolution. It is concerned with macro-evolutionary phenomena such as the pace of change and diversification in material culture. A central premise of this paper is that much of what Palaeoanthropologists take as evidence for innovation emerges from social and demographic phenomena as much as, if not more than, from cognitive ones. It is not an account of any particular interval of time or any special products of human genius; instead, it focuses on the factors which might affect the rate at which new things (artefacts, processes, etc.) make their appearance in the record. This sort of emergent, aggregate creativity is a product of two broad sets of influences. One consists of the biologically based cognitive capacities that affect the abilities of hominins to produce new things, to come up with novel solutions to problems. Once invented, however, new ideas and ways of doing things must also take root. Unless innovations are widely adopted, unless they become commonplace and traditional, the prehistorian is unlikely to recognise them in archaeological contexts. A second set of influences therefore involves the factors that foster the wider diffusion and persistence of that which the individual brain invents.

6.1.1. Variable Pace of Innovation in the Palaeoanthropological Record

By virtue of their subject matter and the record with which they work, researchers of human evolution have a distinct perspective on creativity and innovation. Early products of individual genius are both uncommon and very limited in their distribution. Prior to about 70,000 years ago, there were very few objects that stood out as emblematic of the individual creative process. Even after this time such objects are far from ubiquitous. For example, during the Upper Palaeolithic (ca. 45–12 ka), iconic phenomena such as cave paintings and finely decorated tools are – with some notable exceptions – confined to limited pockets within Western, Central and Eastern Europe. These phenomena constitute a fascinating and rich subject for research (addressed by other papers in this volume), but concentrating too much attention on them leaves out most of the globe and the largest part of human prehistory. If we were to adopt a long-term evolutionary perspective on human innovation, we need to find forms of evidence that provide a continuous and widespread record rather than things which only appear on one corner of the evolutionary stage very near to the end of the play.

The most ubiquitous evidence for human action and expression prior to 70 ka, and after it in most places, comes from stone tools. Stone tools were used everywhere and they preserve extremely well. These products are often unremarkable, usually mundane and almost always functional in nature, but they are irreplaceable documents of the capacities of hominins to innovate. Earlier generations of prehistorians worked mainly on documenting the shapes of stone artefacts; however, in the past few years, researchers have focused increasingly on procedures and techniques used for their manufacture. The study of *chaînes opératoires*, essentially the organised sequences of actions involved in the creation, use and maintenance of artefacts (Boëda, 1991; Boëda et al., 1990; Pelegrin, 1990; Lemonnier, 1992), has provided a remarkably rich field for describing the behaviour of ancient toolmakers.

In the absence of spectacular works of art or craftsmanship, *variety* or *diversity* of material products created by past humans is a crucial source of information about hominins' abilities and propensities for innovation. Variability and change in material culture are *prima facie* evidence of some kind of creative process. The probability that a Palaeolithic archaeologist identifies the spot where some tool form or process was first invented is vanishingly small. However, the existence of diversity requires a source of novelty, and, in the end, what else is creativity but the production of novel kinds of things? It may not be the flashing, mercurial creative act of an artisan: it takes place at a very different timescale, and is controlled by different mechanisms. However, it is a form of innovation nonetheless. Evidence suggests that the potential for creativity was not evenly distributed among early humans. Beginning with the earliest production of artefacts, which currently dates to about 2.6 my, the rates of innovation seem to vary both spatially and chronologically. In other words, there is much variability to be explained.

It is almost a truism that material culture became more complex and diverse over time. It is also quite apparent that rates of technological change accelerated over the course of human evolution, although there are substantial methodological hurdles to measuring this (see review in Stout, 2011). The earliest technologies, dating to between roughly 2.6 ma and 1.0 ma, appear to change very slowly (Ambrose, 2001; Foley and Lahr, 2003; de la Torre, 2011). New techniques or artefact forms appear only at intervals of hundreds of thousands of years. The rate of change in material culture in these early time frames is hardly greater than the rate at which the hominin skeleton evolved, such that there are close correlations between some hominin forms and specific technological modes (Foley and Lahr, 2003). Rates of turnover and diversification seem to increase over the course of the Palaeolithic, but change remains very gradual compared to what we know from the last few millennia. Even as late as the Middle Palaeolithic (250 ka–35 ka), the time of Neanderthals in Eurasia, many researchers are struck by the persistent lack of novelty in technological behaviour compared to what is known slightly later in the Upper Palaeolithic (e.g., Foley and Lahr, 2003: 119; Wynn and Coolidge, 2004; Hoffecker, 2005; Bar-Yosef, 2006: 16).

Although stone artefacts are the most common, best-studied and most readily quantified sources of evidence about the behaviour of ancient humans and human ancestors, they are not the only ones. Evidence for use of fire is more scattered and less evenly preserved than stone tools, but it nonetheless attests to acceleration in the generation of novelty. Hominins may have first begun to use fire more than 780,000 years ago (Goren-Inbar et al., 2004), if not earlier (Brain and Sillen, 1988; Wrangham, 2009). By 400,000 years ago, fire was controlled well enough to be a regular part of daily activities (e.g., Karkanas et al., 2007; Berna and Goldberg, 2008); thereafter, pyrotechnology becomes increasingly diversified. New ways of containing and feeding fires as sources of heat and light appeared (e.g., Rigaud et al., 1995; Meignen et al., 2001; Karkanas et al., 2004). Eventually, fire itself became a tool that Palaeolithic hominins used to change the properties of other technological materials (e.g., Vandiver et al., 1989; Mazza et al., 2006; Brown et al., 2009).

Somewhat less direct clues about emergent, aggregate forms of creativity come from the environmental contexts in which artefacts and human remains are found. Sometime around 2 ma, if not earlier, the first hominins ventured outside of the topical African heartland where they had first evolved (see recent summary of evidence in Dennel, 2009: 186–202). The first forays into the temperate zones occurred soon after, although these were probably confined to fairly warm interglacial periods. By around 500 ka hominin populations were well established in the mid-latitudes (Roebroeks, 2006). Ranges continued to expand thereafter, reaching unglaciated parts of Northern and Eastern Europe by the Middle Palaeolithic, and even farther north after modern *H. sapiens* populations came on the scene (Gamble, 1996; Hoffecker, 2005). The success of Lower and Middle Pleistocene hominins in colonising highly seasonal environments unlike anything experienced by their tropical ancestors implies an ability to cope with new kinds of animal, vegetal and mineral resources as well as novel climatic conditions: in short, it shows an ability to develop new ways of doing things, to innovate.

Although they are valid at a very broad level, models of gradual knowledge accretion and incremental improvements in technological efficiency do not tell the whole story. Long-term trends in the appearance of novelty and diversification during the Palaeolithic are not monotonic. The pace of technological change does not accelerate smoothly. Prolonged periods of apparent stasis in behaviour may be punctuated by periods of seemingly abrupt change. The "creative explosions" of the African late Middle Stone Age and European Upper Palaeolithic, which produced the world's first canons of figurative art and ornamentation, are widely considered watershed events in human evolutionary history. Whether they consider them true explosions or the cumulative result of a long, slow burn (e.g., McBrearty and Brooks, 2000), most researchers would identify these as intervals of heightened creativity. Both of these "events" are thought by many researchers to be linked to the emergence and dispersal of *H. sapiens*, although there is increasing evidence that Neanderthals played a role in the Upper Palaeolithic "revolution" (reviewed by d'Errico and Stringer, 2011). However, periods of change are not usually associated with the appearance of new hominin species. Researchers have begun to concern themselves with trends in cultural evolution that are independent of these important biological transitions (e.g., papers in Hovers and Kuhn, 2005). The late Middle Palaeolithic in Europe, associated with Neanderthals, may have seen a marked upward trend in the pace of change in material culture, which did not necessarily anticipate the succeeding Upper Palaeolithic (Bolus and Conard, 2001; Kuhn, 2005; Soressi, 2006; Slimak, 2008). Another interval of marked diversification in technology occurred during the later part of the Middle Pleistocene in the Near East. Beginning around 400 ka a relatively monotonous late Acheulean was supplanted by a series of archaeological assemblages notable for the variety of technological procedures represented (e.g., Jelinek, 1990; Copeland, 2000; Vishnyatsky, 2000; Barkai et al., 2009). Likewise, the process of habitat expansion was not continuous but proceeded in a stepwise fashion. There were long time delays in penetrating certain habitats, such as the lag in colonising Central and Northern Europe (Roebroeks, 2006).

In fact, general trends towards accelerated change and increasing diversity can be interrupted or even reversed locally. As will be discussed in greater detail later, the earliest Upper Palaeolithic cultures in Europe, dating from ca. 45 ka to 35 ka, show a notable level of geographic diversity. However, the succeeding period is characterised by two very widespread archaeological cultures, the Aurignacian and Gravettian, which are associated with an equally notable continuity in material culture over very large areas. Likewise, the diversity of assemblages in the Near East after 400 ka was truncated abruptly around 250 ka: for the next 200,000 years, a comparatively homogeneous set of assemblages termed Levallois Mousterian are found across the region (Meignen, 2007). In southern Africa, the impressive cultural developments of the later Middle Stone Age, Stillbay and Howieson's Poort cultures (roughly 77 to 58 ka by the most recent estimates (Jacobs et al., 2008; Jacobs and Roberts, 2008)), were replaced by a "post-Howieson's Poort" MSA (Deacon, 1989; McCall, 2007; Lombard and Parsons, 2010), which retains few of the novel forms of material culture and in fact resembles much older Middle Pleistocene MSA variants.

Finally, factors influencing aggregate, emergent forms of innovation and creativity were not evenly distributed within hominin species or during particular intervals of time. Trajectories of development in Palaeolithic technologies proceeded along independent pathways, and apparently at different rates, in different regions. Perhaps most famously, the pace and direction of technological evolution in East Asia during the Lower and Middle Palaeolithic periods were radically different from western Asia, Europe and Africa. Many of the key technological developments seen in Europe and Africa – such as bifacial handaxes and Levallois method – seem to be scarce or absent in the East Asian early Palaeolithic. When they are present, they occur in very different forms (Dennell, 2009: 434). Originally interpreted as evidence for differences in the cognitive capacities and social evolutionary development of hominins (Movius, 1948), the contrasts between East and West are now seen as simply representing alternative evolutionary pathways (Dennell, 2009:433–437; Boëda and Hou, 2010). There are less spectacular regional differences as well. For example, during OIS stages 4 and 3 (ca. 75 ka to 35 ka), Mousterian sites in southwest France show considerable diversity in the forms of artefacts and procedures or producing them (Delagnes and Meignen, 2005; Meignen et al., 2009). Mousterian assemblages created during this same interval in Northern Iberia, one of the few regions with a similar density of sites, seem much less diverse and differentiated (Casanova I Martí et al., 2009).

6.1.2. Innovation and Creativity as Emergent Phenomena

Although Palaeolithic archaeology is an object-focused discipline, individual artefacts (as opposed to individual hominin fossils) generally carry little weight in arguments about evolutionary processes. The archaeological record of early humans and human ancestors is dominated by large numbers of mundane, utilitarian things. The principal units of study are assemblages of artefacts or other material remains. These are collections of material produced and deposited by unknown numbers of makers over periods ranging from decades to centuries or millennia. A singular object or the first example of a new artefact form is likely to go unrecognised, hidden within a much larger background of unremarkable and redundant artefacts. Even when they can be picked out of the background, it is often difficult to know what to make of unique specimens (e.g., see d'Errico and Nowell, 2000). Palaeolithic archaeologists are much more likely to recognise invention or novelty when it becomes widespread, when it makes the transition from new idea to common practice.

Although he works on very different time periods and uses very different kinds of evidence (manly historical records), M. Schiffer makes the key observation that the appearance and persistence of a novel technology require much more than simple invention. Following other historians of technology, Schiffer (1992, 2002) postulates a series of steps that must occur for a community of people to adopt a new way of doing things. These include experimentation with and refining the processes or forms for local circumstances, and, most crucially, the copying of the novel behaviour beyond a small sphere of specialists or early adopters. For a new behaviour to become widespread and persistent, it is essential that many people learn about it and find it useful. Somewhat paradoxically, originality on the part of a single individual or one community becomes visible only by virtue of a profound lack of originality on the part of many others (Kuhn and Stiner, 1998): in effect, "…social learning parasitises innovation" (McElreath, 2010: 456). The initial spread and subsequent persistence of an original artefact form or novel procedure thus depend most fundamentally on processes and conditions that allow people to learn new ways of doing things and to make decisions about whether to emulate them.

A range of factors can contribute to this sort of emergent creativity. These can be divided roughly into factors that influence the production of novelty and those that influence the spread, persistence and disappearance of that novelty. Perhaps the most obvious influence on rates of innovation stems from the cognitive processes underlying the creative act. Hominin cognition no doubt has changed in radical ways since our lineage split from that leading to modern chimpanzees and bonobos seven million years ago, and no one would deny that humans' potential for imagination and invention has grown enormously over their evolutionary history. Consequently, the explanation of first resort for long-term trends in the material expressions of creativity and innovation normally involves biologically underwritten changes in human cognitive capacities.

At the same time, cognitive evolution cannot explain all of the variability in rates of change and diversification described above. Gradual increases in rates of behavioural change and diversification may well demonstrate species-wide cognitive developments, but cognitive differences would not explain differences in evolutionary trajectories between East Asia and western Europe, nor would they account for different levels of technological diversity among contemporaneous Neanderthal populations in Spain and southwest France. Intervals of accelerated change followed by comparative stasis, or even trend reversals, are also unlikely to map directly onto changes in hominins' inherent capacities for producing new things.

In fact, the importance of innovation may be over-stressed in accounts of human cultural evolution (Henrich et al., 2008: 129–131), especially when dealing with long-term records. An implicit, and sometimes explicit, distinction is drawn between mutation and innovation. The former generates novelty in an essentially random manner, whereas the latter is seen as purposive and goal directed. Even if we accept the importance of goal-directed problem-solving behaviour as a motor for innovation in the comparatively brief frame of recent history (but see Basalla, 1988; Meyers, 2007), its importance to the kinds of emergent patterns studied by Palaeolithic archaeologists remains undemonstrated. Simple errors in copying established procedures and forms, analogous to genetic "drift", would produce novel ways of doing things just as does directed innovation. It does not necessarily matter how novel forms of behaviour originated: what matters is whether they spread and persisted. Particularly when dealing with large-scale, aggregate trends we may be justified in treating random and purposive sources of novel behaviours as equivalent. Random processes of genetic mutation, winnowed by natural selection, can provide apparent direction to biological evolution. Why should we not assume that mistakes in replicating cultural information would have similar consequences?

The converse is also true. The absence of clear directional trends or strong biases in the aggregate does not necessarily mean that there was no goal-biased behaviour at the individual level. Individual behaviours that are highly constrained and tightly determined may appear to behave stochastically at the group level. Among a host of recent examples are the frequencies of baby names in birth records (Hahn and Bentley, 2003; Bentley et al., 2004). Few parents select their child's names at random: they have strong preferences for particular names and against others. However, the distribution of these preferences is such that the aggregate pattern mimics a stochastic process:

"Practically speaking, the random copying model does not require that people make choices without any reasons at all, but rather that the statistics of all their idiosyncratic choices, at the population level, are comparable to random copying" (Bentley, 2009: 122).

The same applies to the behaviour of early humans leading to the accumulation of archaeological assemblages. Individual artisans were certainly trying to solve very immediate problems, but not everyone necessarily faced the same problems. It is the aggregate distribution of different kinds of problems that determined the macroscopic patterns we perceive as the distribution of different technological forms or rates of variation through time and across space.

Forces of population genetics are more important than mutation in affecting macro-evolutionary change within species. In the same way, the group-level processes that lead to spread, imitation, persistence and loss of novelty are certainly as important as, and arguably more important than, the cognitive sources of new ideas in determining the broad trends and tendencies that palaeoanthropologists take as evidence for creativity. Whether we think of innovation as random or goal directed, Schiffer and others (e.g., Rogers, 1983) have shown that a whole series of social and cultural processes must intervene for change to be detectable at the level of communities, societies or populations. Many of the factors that Schiffer and Rogers identify in industrial societies clearly do not apply to the Palaeolithic. Nonetheless, some minimal conditions must be met for an advantageous innovation to spread. Most importantly, a range of individuals must be exposed to the innovation, must recognise its value and must be free to change their ways of doing things. In the case of novel behaviours with little immediate advantage over existing ones (so-called neutral variants), only the frequency of exposure and rules for copying determine whether a particular variant will become established in a population. In nonliterate societies where communication is primarily face to face, multiple or prolonged exposures are probably required for a new idea "takes hold." To leave a recognisable signal in the Palaeoanthropological record, novel behaviours must also persist and must be passed on from one generation to the next with a certain degree of fidelity.

I do not mean to argue that goal-directed behaviours are inconsequential as regards long-term patterns of culture change in the Palaeolithic. We must assume that most acts which produced an artefact or other trace of behaviour had a purpose. However, given the periods of time (chronological and generational) sampled by assemblages of artefacts, the minimal units of analysis in most studies of the Palaeolithic, we can also assume that they encompass purposive behaviour directed at *a great many* more-or-less independent aims. The intentionality implied by the production of a single artefact is often lost in the emergent, aggregate properties of assemblages, archaeological sequences and larger geographic groupings of things.

In some respects, it is useful to think of emergent, aggregate patterns of behaviour as analogous to a flock of birds or swarm of bees. Ethological and experimental studies have shown that a swarm may move in a very coherent and unified manner, even if none of the members has the aim of making it do so. Rather, the properties of the whole emerge from interactions between individuals guided by simple rules. The entire group may even appear to act in consensus based on information that only a small fraction of the members actually possess (Beekman et al., 2006; Seeley et al., 2006). Clearly, the motivations and behavioural rules of humans and human ancestors are orders of magnitude more diverse and more complex than those of bees. This fact notwithstanding, the rules that allow the group as a whole to exhibit coherent, directional behaviour may be quite simple. An emerging field of enquiry, dual-inheritance theory, is directed at understanding how simple rules of experimentation and imitation may lead to complex, emergent group dynamics (Boyd and Richerson, 1985, 2005) of the type which give rise to macroscopic patterns of culture change and diversification.

A complete account of fluctuating rates of change and varying levels of diversity in early hominin behaviour ultimately requires integrating cognitive evolution with an understanding of the mechanisms affecting the diffusion and transmission of novel ideas *among* people. In the remainder of this paper, I will focus mainly on the second set of factors. This is not to deny the importance of cognitive evolution: it is more a function of limitations in space and in my own training. Moreover, others are much better qualified to discuss the evolution of the human mind (e.g., Mithen, 1996; Noble and Davidson, 1996; Dunbar, 1996; Carruthers and Chamberain, 2002; Klein and Edgar, 2002; Beaune et al., 2009; Coolidge and Wynn, 2009; to cite just a few). I first review recent work on the consequences of population size for the diffusion and retention of novel ways of doing things. Recent work, drawing on principles from population genetics and dual-inheritance theory, demonstrate how demographic factors, including sizes of standing populations and rates of extinction, could affect diversity and rates of change in material culture. I will then turn to the potential effects of social network structures on the spread of new ideas and procedures. Degrees and patterns of connectedness or network structure in ancient populations would likewise have had profound consequences for the diffusion and retention of novel things. The final section briefly addresses some consequences of interactions between demographic factors and innovation at the individual level.

6.1.3. The Effects of Population Size and Robusticity on Behavioural Change and Diversity

Over the past decade, a number of scientists have examined the effects of population size and structure on diversity and change in material culture, an area of enquiry cleverly labelled "evo-demo" (Richerson et al., 2009: 214). One of the pioneering works in this field is Shennan's (2001) study, which uses a modified version of the Peck et al. (1997) simulation model to explore the effects of population size on mean fitness under conditions where culturally transmitted traits had direct fitness consequences. In this model, "innovations" occur randomly and are thus analogous to biological mutations: a novel behaviour is as likely to result in decreased fitness as in a fitness increase. Shennan is able to show that for a range of seemingly reasonable parameter settings, the size of the effective population is directly correlated with average fitness. This is true for both vertical (parent–offspring) and oblique (age-biased but not familial) transmission of cultural traits. The explanation is fairly straightforward. In larger populations, innovations with high fitness values are more likely to arise and subsequently be adopted by others. A similar argument has been made concerning rates of genetic evolution in humans since the Neolithic (Hawks et al., 2007). In a subsequent paper, Powell et al. (2009) use a more sophisticated model developed by Henrich (2004) to further explore this phenomenon. They conclude that a range of demographic factors, including both density and intergroup migration, can directly influence the accumulation and maintenance of complex cultural skills. These skills, in turn, enhance the success of a population, creating a positive feedback situation (see also Richerson et al., 2009). In a study of innovation rates and the sizes of cities, Bettencourt et al. (2007) show that similar processes operate in the contemporary world.

The papers by Shennan (2001) and Powell et al. (2009) both go on to discuss intervals of accelerated cultural change and diversification associated with the early Upper Palaeolithic in western Eurasia and the late Middle Stone Age in Southern Africa. Based on a review of evidence from genetics, radiocarbon dates and archaeology, they argue that these two instances of cultural fluorescence in the late Pleistocene were in fact associated with demographic expansion or increasing population density. In part, these arguments were proposed as alternatives to explanations based on putative changes in biologically based cognitive capacities linked to the appearance of *H. sapiens* (e.g., Mithen, 1996; Klein, 2000). However, demographically based models of cultural evolution cannot disconfirm models of cognitive evolution, and vice versa: they simply establish that demography can influence the rates at which new and beneficial innovations appear and spread. Ultimately, the strength of either explanation rests on independent evidence for evolutionary changes in cognitive abilities or population sizes.

Although demographically based models have been proposed to account for periods of accelerated change and diversification in the early Upper Palaeolithic and later Middle Stone Age (see also d'Errico and Stringer, 2011; Straus, 2010), they are equally applicable to other periods and places. One possible example is the contrast between the late Middle Palaeolithic of France and Iberia. The fact that roughly contemporary populations of Neanderthals in neighbouring regions seem to have exhibited different levels of technical and cultural diversity begs an explanation that is not founded on differences in fundamental abilities. Speaking more broadly, we might predict that rates of cultural change and diversification would be correlated with environmental variables such as temperature and moisture availability if these factors constrained the sizes of hominin populations.

Populations can shrink as well as grow, and demographic declines can also have significant impacts on cultural diversity and the potential for change. Henrich (2004) explores the effects of population loss or very small population sizes in prehistoric Tasmania on the accumulation and retention of cultural skills. He demonstrates that rapid shrinking of populations accompanied by isolation from larger pools of cultural information on the Australian mainland could explain the loss of cultural traits (but see Read, 2006 for a critique). This is more than a drift-like process affecting behaviours with little or no adaptive value. A significant decline in population size can also lead to the loss of advantageous forms of behaviour, especially if these behaviours are complex, difficult to learn and, hence, normally are acquired by only a limited subset of the population (Henrich, 2004: 203–205). Given the likelihood for significant contractions in Palaeolithic populations in the face of major environmental perturbations such as Heinrich events (abrupt, rapid-onset cold snaps) (Heinrich, 1988; Maslin et al., 2001) and volcanic eruptions (Fedele et al., 2008; Golovanona et al., 2010), it would be interesting to know whether severe environmental perturbations were associated with contractions in cultural diversity or intervals of depressed change. Interestingly, diversity of Middle Palaeolithic toolkits, as measured using the standard Bordes (1961) typology, shows no clear correlations with glacial and interglacial cycles in Pleistocene Europe (Bocquet-Appel and Tuffreau, 2009). However, it is not clear that this particular dimension of variation is telling us something important about Neanderthal culture: the typological system may simply be an inappropriate instrument for measuring variation in artefact function or design.

Lycett and Norton (2010) have drawn on Henrich's model to argue that the apparent "impoverishment" of East Asian early Palaeolithic technologies compared to those of Europe and Africa could be a function of smaller absolute population sizes. It has long been observed that many developments in Palaeolithic technology, such as bifacially shaped handaxes and Levallois flake production, never appear east of India during the Lower or Middle Palaeolithic. In fact, recent evidence suggests that while handaxes at least were present, they were of a locally evolved, simple form (Dennel, 2009: 433–437). Lycett and Norton make the case that what is "missing" from the East Asian record are those procedures that require the greatest knowledge and skill to execute. Following Henrich's lead, they argue that these would be exactly the kinds of practices most likely to disappear as colonising populations underwent major bottlenecks. Meanwhile, larger and better-established populations in Africa would have retained this basic knowledge by virtue of a larger, more stable population base.

Another possible example of the negative effects of population sizes on behavioural diversity comes from the earliest occupation of Europe. Over the past two decades, a series of finds have established conclusively that hominins were present at the southern and eastern edges of Europe by 1 mya, if not long before then. These early sites are sparsely distributed and often quite "thin" in the sense that artefacts are rare relative to palaeontological specimens, which in turn suggests that hominin populations were small. Interestingly, the artefacts found in these early sites are quite simple flakes and cores (e.g., Carbonell et al., 1999; Arzarello et al., 2007; Güleç et al., 2009; de Lumley et al., 2010), similar to the Chinese Lower Palaeolithic and the Oldowan of Africa. The contents of even rarer sites in Northern Europe dating to before 500 ka are similar (e.g., Parfitt et al., 2005). Meanwhile, contemporaneous sites in North Africa and, more proximately, in the Levant (Bar-Yosef and Goren-Inbar, 1993) contain Acheulean assemblages with more complex, bifacially worked tools and an evidence of a greater variety of technological procedures.

We would not want to attribute the "impoverishment" of the earliest Lower Palaeolithic artefact assemblages from Europe to a lack of mental prowess on the part of their makers. After all, these same populations were pushing into new habitats previously unoccupied by hominins, implying a certain level of behavioural flexibility and the ability to deal with novel resources and conditions. However, the presence of these very simple technologies could be an instance where some more rarified kinds of technological knowledge were lost as a function of the obstacles faced by tropical hominins in colonising the temperate zones. The uncertainties of living at the extreme edges of their species range could have made for small and unstable hominin populations early on, just the kind of situation in which cultural knowledge could easily be lost by chance events. It is not until much later, after hominins had solved some of the problems of survival in seasonal temperate environments and established archeologically visible populations in mid-latitude Europe (Roebroeks, 2006), that the more elaborate kinds of technologies such as typical Acheulean bifacial manufacture also became prevalent (Santoja and Villa, 2006; Villa, 2009: 265–266).

The sizes of hominin populations do not tell the whole story. Population stability may be another contributor to long-term trajectories of change and diversification in cultural behaviour. There are many reasons to suspect that hunter–gatherer populations, and indeed later ones as well, underwent cycles growth and decline, with periods of local extinction (Boone, 2002; Shennan and Edinborough, 2007; Stiner and Kuhn, 2006; Riede, 2008). Premo and Kuhn (2010) used a spatially explicit agent-based model to examine whether rates of extinction in local subpopulations could influence the appearance of culture change and diversification over the long term. Their model showed how increasing rates of local extinction could depress cumulative change, total diversity and intergroup variation, even when overall meta-population sizes remained constant. Periods of apparently slow behavioural evolution, characteristic of much of the record prior to (and arguably after) the origins of *H. sapiens*, could stem partly from the fragility of local populations. By implication, periods of accelerated change and diversification could reflect greater continuity and robustness of local populations in the face of environmental fluctuations.

Models of cultural transmission and population size and density provide novel and potentially productive explanations for periods of rapid technological change as well as stasis or impoverishment in the Palaeolithic record. There are, however, many questions and problems yet to be resolved. For example, the effects of population size and density cut both ways. Depending on the scale of analysis and the particular characters in question, upward and downward fluctuations in population could have similar effects. In theory, small, sparse populations could result in the loss of rare behaviours or unevenly distributed knowledge, even if the knowledge has adaptive value. However, small populations isolated from one another could well loose, or preserve, different variants through drift-like processes, resulting in greater geographic diversity. Larger, denser populations meanwhile can foster the development of a greater variety of behavioural options. Especially advantageous variants that happen to appear can spread quickly within large, dense populations, replacing less effective behaviours. A case in point may be the rapid spread of microlithic technologies in India during the later Pleistocene (OIS 3), which appears to coincide with a period of increased population and shrinking territories (Petraglia et al., 2009). While this situation would lead to the appearance of high rates of innovation and rapid change, in the long run it could also result in reduced variability. The key to resolving this kind of equifinality is careful attention to variation at different geographic scales (local and regional), as well as a concerted attempt to distinguish neutral behavioural characters from ones with energetic or fitness advantages. Principles from the neutral theory of biogeography (Hubbell, 2001), and particularly as regards the relationship between measures of diversity at different spatial scales, may be important tools for working out the effects of population sizes on Palaeolithic material culture.

6.1.4. The Effects of Connectivity and Networks

From the perspective of cultural transmission, the sizes and densities of populations cannot be measured just by counting the number of individuals living in a particular area. The size of the group that may be exposed to a new idea is also a function of social networks that link individuals and local groups within a larger population (Shennan, 2001: 12; Stiner and Kuhn, 2006). In a regional meta-population of hominins that is partitioned into small local groups that are effectively isolated from one another the effective local population sizes for cultural (and reproductive) purposes are much smaller than the census population for the species: only a small fraction of the total population ever exchanges ideas or genes. If, on the other hand, the local groups are in effective contact with each other and novel ideas can travel among them, then the effective population size can be as large as the census population. In other words, changes in social networks, connections among individuals and groups, may have the same kinds of effects as total population sizes (see also Powell et al., 2009).

Humans are certainly avid networkers, and we have been since well before the invention of Facebook®. Anthropologists and sociologists have been interested in the structures created by ties among individuals for as long as the disciplines have existed. More recently, mathematical theorists have identified a number of different network topologies or structures relevant to human and animal societies (e.g., Newman et al., 2006; Borgatti et al., 2009). The density and structure of connections among nodes (individuals) can influence the rate at which information (or anything else) propagates across geographic and social space. As such, the formation of social networks and their structures could have important consequences for the dispersal and retention of novel behaviours, and so for the appearance of innovation in the Palaeolithic cultural record.

Several factors may influence the density and structure of social connections within human populations. All other things being equal, population density should be directly correlated with the intensity of social interactions among social groups. Even if connections are formed randomly, members of denser populations should bump into one another more frequently. The physical structure of environments, particularly the existence of geographic barriers, is another obvious influence. Less obvious factors are environmental richness and the subsistence choices made by prehistoric foragers. Scales of logistical mobility and the sizes of foraging radii, the distances that foragers move out from residential camps, directly influence the likelihood that individuals will encounter each other on the landscape, and therefore the frequency of opportunities to exchange ideas and techniques (Perreault and Brantingham, 2010). Scales of mobility are in turn affected by environmental quality and the kinds of resources that foragers target (Kelly, 1983, 1995; Binford, 2001; Hamilton et al., 2007; Grove, 2009). The less dense the distribution of resources, the larger the territories foragers have to cover. Likewise, hunter-gatherers that depend on thinly distributed resources such as large game should move over larger areas than foragers with economies based on more densely concentrated foods such as fish or patchy vegetable resources. Thus, the nature of hominins' habitats and diets could have direct consequences on the rates that individuals from different local groups encountered one another, and the likelihood that they would exchange cultural information.

If scales of mobility could affect the formation of social networks and the potential for the spread and retention of novel cultural traits on them, it is worthwhile asking when major changes in territoriality and land use might have occurred over the course of human evolution. We can assume that the first hominins, like our closest relatives, the chimpanzees and bonobos, lived in social groups knit by intense interactions among group members. At some point, their social lives began to expand far beyond the local group. Hominins first began to systematically exploit large animals as food sources, whether by hunting or scavenging, around 2.5 mya (reviewed by Plummer, 2004; but see McPherron et al., 2010 for possible evidence of even earlier forays into carnivory). On purely theoretical grounds, we might expect expansion of territories beyond patches of forest to have accompanied this inclusion of more meat into the diet. Anatomical evidence from later *Homo ergaster/erectus* (after 1.8 my) suggests an adaptation to a high level of mobility, particularly long-distance running (Bramble and Lieberman, 2004). Both of these behavioural trends are linked with well-recognised technological developments. The first widespread evidence of carnivory around 2.5 my is of course associated with the earliest stone tools (Plummer, 2004; de la Torre, 2011). Anatomical developments linked to higher mobility in *H. erectus* coincide roughly with the widespread appearance of the earliest bifacial (Mode 2 or Acheulean) stone artefact technologies, which in theory at least involved a more complex and demanding mode of manufacture than Oldowan technologies (Coolidge and Wynn, 2009: 165–170; Faisal et al., 2010; Stout, 2011). There may be basic functional explanations for both the first development of stone tools and the subsequent shift to bifacial shaping of artefacts. However, the dispersal of these new behaviours across large parts of the African continent, and eventually into Eurasia, required at least some exchange of information among local populations.

The influence of environmental factors on the propagation of novel ideas would be strongest where social networks were determined by random encounters. Recent humans at least are not content to form ties based on chance meetings; instead, they actively seek to form relationships with some individuals but not with others. Strategies for creating ties have profound effects on the number and structure of connections within social networks and on the potential for cultural information to spread and persist within them.

A particularly important sort of network structure for the study of human societies is "small world" networks. Small-world networks (Watts and Strogatz, 1998) are characterised by a high degree of connectedness within local groups and sparse but consistent ties among distant groups. In a small-world network, members of a local clique, subpopulation or residential group would all tend to have close ties to one another. At the same time, most individual members also have a few friends or allies in more distant groups. These long-distance links have important effects. The existence of even a few interconnections among distant individuals in such network structures tends to minimise the average path lengths between any given pair of individuals. This phenomenon, also described as the "strength of weak ties" (Granovetter, 1973), is thought to be behind the widely discussed "six degrees of separation" among people in the United States, the observation that, on average, a chain of no more than six individual acquaintances separates any two people in America. More important for the current discussion is that a network structure that minimises average social distances will maximise the rates at which "contagions" such as novel cultural behaviours can spread (Watts and Strogatz, 1998), whether by random copying or selective emulation. This has been confirmed by simulation for simple traits (Cowan and Jonard, 2004), although it may be less effective for complex behaviours that require information from multiple sources (Centola and Macy, 2007).

Network structures can also help to mitigate the effects of local extinction on culture change and diversity. Building on an earlier paper (Premo and Kuhn, 2010), Premo asked whether the exchange of ideas between groups could counteract the tendency of high rates of extinction to depress both change and variation. He discovered that by facilitating the spread of traits over larger areas a wide range of network structures could effectively buffer the influence that local extinctions could otherwise have on cultural diversity (Premo, submitted for publication).

Small-world topologies can facilitate the rapid spread of ideas. They are also likely to characterise the social networks of Palaeolithic and later hunter-gatherers. Most hunter-gatherers live in small bands that change residential locations periodically. Relationships among band members are close: in a face-to-face society, everyone knows everyone else. Even so, the social lives of hunter-gatherers extend well beyond their local band. Adults often maintain ties with individuals in other groups. Sometimes these are bonds of kinship: foragers in arid Australia are well known for their complex and geographically vast kinship networks (Yengoyan, 1979). However, members of hunter–gatherer societies may also be tied to distant individuals who are not kin. Phenomena such as the Hxaro system among Khoi San foragers in southern Africa (Wiessner, 1977, 1982; Schweizer, 1996) or trade partnerships among Inuit groups (Burch, 1991) are well-known examples. In these systems, individuals go out of their way to establish bonds with partners in distant areas, cementing the relationship through the exchange of gifts. The social ties that they create do more than just provide access to goods from distant sources. For one thing, they allow individuals to obtain information about resources and social conditions in far-flung locations (Whallon, 2006). More importantly, they can serve as a kind of social insurance. Individuals or their families may take advantage of bonds of friendship to move into a partner's foraging territory when conditions in their own area are bad (e.g., Yengoyan, 1979: 399–400; Burch, 1991: 104; Kelly, 1995: 188; Binford, 2001: 466).

There are many benefits to establishing these sorts of long-distance ties irrespective of their consequences for information transfer. However, once established these long-distance ties would naturally facilitate the spread of innovations across human populations.

Although there has been a great deal of recent theoretical work on network structures in the abstract and in the modern world (e.g., Newman et al., 2006), there are not many formal studies of network structures among foragers, much less among Palaeolithic hunter-gatherers (but see Schweizer, 1996). Attempts to test the proposed relationship between population sizes and diversity of cultural behaviours nonetheless may provide indirect evidence of how intergroup connections foster or repress cultural diversity. Using ethnographic data, Kline and Boyd (2010) show that population size is a better predictor of the diversity of artefact forms used by groups in Oceania than any environmental factor alone. They also observed that the societies in their sample which enjoyed high levels of intergroup contact also had more diverse technological repertoires than expected. The more isolated groups had fewer types of artefacts than would be predicted by population size alone. Although previous studies by Collard et al. (2005) and Read (2008) failed to find a relationship between population size and diversity in artefact forms, Kline and Boyd attribute this to the fact that the other researchers examined ethnographic groups in western North America known to be in close contact with one another. In other words, the movement of ideas among groups on established social networks ameliorated the effects of different population sizes (Kline and Boyd, 2010: 2560).

Direct evidence for large-scale social networks is strongest for the later periods of human history, starting in the late Middle Stone Age and Upper Palaeolithic. Social ties among foragers living in different residential groups are often cemented by the exchange of goods: sometimes these objects are of intrinsic value, but often they are more important as tokens of relationships among individuals. Movement of items such as stone raw materials or shells from identifiable sources can help in reconstructing the scales and even the structures of social networks. In Eurasia, evidence for displacement of materials over long distances (>40–50 km) is rare prior to the Upper Palaeolithic. Middle Palaeolithic assemblages contain a narrow array of material culture forms, overwhelmingly dominated by stone tools. The great majority of artefacts are local in the sense that they are made of raw materials collected within 20 km or so of a site. There is occasional evidence for displacement of artefacts over distances exceeding 100 km during the Middle Palaeolithic (Feblot-Augustins, 2009; Meignen et al., 2009; Kuhn, 2011). These long-distance "raw material transfers" are uncommon, and normally involve just a few well-used artefacts (e.g., Slimak and Giraud, 2007). It is possible that they are evidence of special exchange relationships, but the rare exotic artefacts may instead be items of transported personal gear that simply remained in use for unusually long periods (Kuhn, 1992; Feblot-Augustins, 2009): as such, they would represent one tail of a continuous range of artefact transport distances (Brantingham, 2003, 2006). Even rarer in Middle Palaeolithic sites are nonutilitarian artefacts such as fragments of marine shells in sites located far inland (Arrizabalaga, 2009). Such occurrences are so infrequent and unique that it is difficult to assess their broader behavioural significance. They could indeed be tokens of exchange relationships, but they could also simply be curiosities that mobile individuals happened to carry along with them in their travels. The sheer difficulty of identifying consistent evidence for exchange of artefacts or raw materials in the Middle Palaeolithic of Eurasia suggests that strategies of social alliance formation were different from, or not as geographically ambitious, as they were among more recent foragers.

Although the density of data is lower than in Europe, the later Middle Stone Age of eastern and southern Africa, associated with early *H. sapiens*, contains clearer traces of long-distance movements of artefacts and raw materials. Current evidence indicates that obsidian – a common object of exchange throughout prehistory – was sometimes moved as far as several hundred kilometres between the source and eventual point of discard (Merrick et al., 1994; Agazi Negash and Shackley, 2006). Although more work is clearly needed, this situation is suggestive of intergroup exchange and a means for forming large networks of social ties.

Somewhat later in time, the Upper Palaeolithic record of Eurasia contains even more abundant evidence for construction of intergroup social networks (Gamble, 1997). Beginning at least with the early Aurignacian, tools and even unfinished artefacts were moved over distances in excess of 100 km on a regular basis (e.g., Feblot-Augustins, 1997, 2009; Bordes et al., 2005; Arrizabalaga, 2009: 52). Artefacts of exotic raw materials even appear in places where good stone is already plentiful, demonstrating that the transport of materials was not intended to remedy local shortfalls. As time goes on, long-distance movement of both utilitarian stone artefacts (Gamble, 1999: 315–320; Feblot-Augustins, 1997, 2009) and nonutilitarian objects such as ornaments (e.g., Hahn, 1972: 261; Taborin, 1993; Gamble, 1999: 321) becomes even more common. The scales over which Upper Palaeolithic artefacts were habitually moved varied across regions, being greater in Eastern and Central Europe than in western Europe (Feblot-Augustins, 1997, 2009). This differential is partly a function of the distribution of flint-bearing geological formations (Duke and Steele, 2009); however, it may also be linked to differences in mobility and strategies of long-distance network formation. We might expect foragers living in harsher, less productive continental habitats to have occupied larger territories (Feblot-Augustins, 1993) and to have actively sought out more geographically extensive alliances as a hedge against local resource failures (Kelly, 1995: 198).

It is reasonable to postulate a causal connection between strategies of social alliance formation and the cultural dynamism of the Eurasian Upper Palaeolithic and late MSA in Africa (Stiner and Kuhn, 2006; Powell et al., 2009). But simply contrasting these long time intervals with earlier periods makes it difficult to disentangle the effects of networks from the influence of demography and the appearance and dispersal of a new hominin, *H. sapiens*; that is, it leaves open the question of whether the rapid innovation during the late MSA and Upper Palaeolithic was due to new kinds of social networks or simply new kinds of hominins. Dynamics within the Upper Palaeolithic may point more directly to the consequences of networks for cultural change and diversity. More specifically, cultural homogenisation and turnover within the Upper Palaeolithic could well be related to the increasing interconnectedness of human populations.

The earliest Upper Palaeolithic of Eurasia, from 45 ka to about 37 ka, shows a high level of regional cultural diversity. "Basal" early Upper Palaeolithic assemblages such as "LRJ" (Lincombian–Ranisian–Jermanowician), Chatelperronian, Uluzzian, Bohunician/Initial Upper Palaeolithic and proto-Aurignacian are identified on the basis of small suites of characteristic artefact forms and technological procedures with limited geographic distributions (Bar-Yosef, 2002; Chabai, 2003; Djindjian et al., 2003; Hoffecker, 2011). At an admittedly coarse scale of observation, a good deal of the variation among these various industries or cultures has no clear functional correlates, except possibly with respect to design of hunting weapons (Bon, 2006; Teyssandier, 2007). Some of these assemblages contain many beads, bone tools and other "hallmarks of modernity", and others do not. Except for early Aurignacian (*H. sapiens*) and Chattelperronian (Neanderthals), we do not know for sure which hominins made them, and even the latter association has been challenged (Bar-Yosef and Bordes, 2010).

This variety in basal UP industries certainly in part represents phyletic diversity. Some basal Upper Palaeolithic industries – such as the Chatelperonian and LRJ – appear to have roots in the local Middle Palaeolithic, whereas others do not. Although two populations of hominins, Neanderthals and anatomically modern *H. sapiens*, are involved, the variation cannot be entirely explained by associations with different hominins. However, all this diversity developed during a period of rapidly shifting environments (Heinrich, 1988; Maslin et al., 2001; Fedele et al., 2008; Golovanona et al., 2010). With two interacting hominin populations experiencing novel demographic and environmental pressures, we should not be surprised to see such a varied cultural patchwork.

What is remarkable is how different the situation looks just a few thousand years later. Around 33 ka, except in a few places where Neanderthals using Middle Palaeolithic (not Upper Palaeolithic) technology were still hanging on, a single culture complex, the Aurignacian, was distributed across much of Europe and even parts of western Asia. The Aurignacian itself is a broad cultural unit with shifting and uncertain boundaries (Bar-Yosef, 2006; Mellars, 2006). Nonetheless, there is evidently much less variation in the design and manufacture of artefacts within the Aurignacian than there was among earlier Upper Palaeolithic assemblages. Replacing the Aurignacian a few thousand years later is the Gravettian, which has an even wider distribution. The near "global" distributions of the later Aurignacian and Gravettian in Eurasia, the presence of single complexes over vast areas that were formerly characterised by substantial diversity, are anomalies to be explained (Zilhão and d'Errico, 2003: 344). Complete population replacements across such a vast area are unlikely, particularly as they would have involved one population of *H. sapiens* replacing another.

It may be more appropriate to think of the broad distribution of the Aurignacian and Gravettian as reflecting their success as cultural phenotypes in dispersing across geographic space. Evidence for expansion in social networks within the Upper Palaeolithic could help explain this dispersal. Whereas the earliest Upper Palaeolithic is quite regionalised technologically and stylistically, the very widespread cultural entities, Aurignacian and Gravettian, that came later were accompanied by expanded evidence for movement of stone artefacts and other objects over substantial distances (Bordes et al., 2005; Bordes, 2006: 157–158; Feblot-Augustins, 2009: Table 3.2). Small-world-like social networks, created quite inadvertently through individual attempts to create social alliances through exchange partnerships, could have enabled rapid diffusion of novel ideas and comparatively rapid turnover of cultures over large areas. This volatility in cultural forms in turn is precisely the kind of thing that Palaeoanthropologists might use to single out periods of heightened creativity or innovation. Larger, more stable populations (Powell et al., 2009; Premo and Kuhn, 2010) could have had additive effects.

It is important to bear in mind that network structures within and among populations can reduce diversity as well as increase it, depending on the circumstances. Densely interconnected social networks increase the effective population size for cultural transmission, thereby reducing the tendency of drift-like processes to eliminate rare cultural variants. At the same time, efficient small-world networks could allow particularly advantageous traits to spread quickly and widely, perhaps displacing even slightly less effective ones. Here again, it is particularly important to distinguish neutral cultural variants from ones that confer some distinct benefits. Focussing on variability expressed at different spatial scales, analogous to α, β and γ diversity in biogeography (Hubbell, 2001), could be an especially useful strategy for isolating the effects of network structures on Palaeolithic material culture.

6.1.5. What About Variation in Innovation Rates?

The models discussed to this point are based on the assumption that individual innovation rates – the frequency with which new things are invented – are independent of demographic conditions and social networks. This may be a consequence of borrowing genetic models, where mutation rates are unaffected by processes operating at the level of populations. However, such an assumption may not be valid for behavioural innovation in human evolution. There is little doubt that humans' capacity to envision and instantiate novel behaviours has increased over our lineage's history, and we certainly expect long-term trends in innovation rates to be linked to the evolution of cognition. More importantly, we should also expect to find short-term fluctuations that reflect pressures and rewards for producing novel products and processes. These rewards for inventing or adopting novel forms of behaviour are in turn quite plausibly linked to Malthusian factors (population/resource balance). To complicate things further, beneficial inventions that lead to improvements in the efficiency or gross productivity of human economic activities should have significant effects on population levels, while network structures may also contribute to their robustness and persistence (Stiner and Kuhn, 2006).

I chose to focus on aggregate rather than individual processes in this paper because the aggregate is what Palaeolithic archaeologists most often study. In fact, cognitive and demographic, individual and aggregate perspectives on innovation should complement one another rather than compete. The parallel development of better models for any of these phenomena should provide sharper, more testable hypotheses for the others. To this end, it is worth discussing some possible relationships between individual cognitive evolution and population-level phenomena in more detail.

McElreath (2010) considers the social conditions that could favour an ability to innovate as well as to acquire cultural information by social learning. He takes a relatively narrow definition of innovation as "…an organism's investment in acquiring new adaptive information directly from the relevant environment…" (McElreath, 2010: 457), in order to distinguish it from social learning or emulation. Consistent with earlier work (summarised in Laland, 2004), McElreath shows that the optimal individual strategy is always a mixture of innovation and emulation, with the optimal mix depending on factors such as the frequency of environmental fluctuations and the number of innovators in the population, among other things. This balancing act does not however lead to global increases in adaptive behaviour, nor does it explain why humans' capacity to invent might have expanded over time. McElreath's solution is to link social learning with innovation: after all, there is no payoff to getting better at acquiring accumulated cultural knowledge unless new knowledge is accumulating at a high rate. If the cost of innovation is slightly higher than that of emulation, the result is a system in which adaptive information accumulates at an accelerating rate (McElreath, 2010: 463) (see also Reader and Laland, 2002). Here lies the link to demography. The availability of beneficial innovations as well as the sheer quantity of accumulated cultural information should both scale with the sizes of populations. We might even anticipate a kind of escalating selective relationship between scales of social interaction and the ability or process and sort through increasing quantities of information obtained by social learning.

Positing large-scale social networks as a contributor to geographic and chronological variability in Palaeolithic behaviour implicates yet another aspect of cognition. A variety of animals, including many primates, live in social groups characterised by intense social interaction. R. Dunbar (Dunbar, 1998; Dunbar and Shultz, 2007) makes a strong case that the unusual development of the brain in higher primates, humans in particular, is a direct consequence of the complexities of social interaction in large groups. Simply increasing the sizes of interacting human populations should place correspondingly greater demands on social intelligence. The habitual formation of long-distance, extra-group bonds might have its own distinct cognitive requirements. Long-distance relationships differ in temporal scale from everyday, face-to-face interactions. Exchange partners might not see each other for weeks, months or years at a time, and so they must maintain their relationship without the benefit of frequent reinforcement. It has been argued that nonhuman primates that live in groups with fluctuating memberships (fission/fusion dynamics) must constantly renegotiate relationships. Not coincidentally they score higher on many tests of intelligence than primates that live in more stable groups (Amici et al., 2008). It is legitimate to ask whether the kinds of social interactions implied by long-distance exchanges and network formation among Palaeolithic groups would impose categorically different demands on cognition than other kinds of social interactions.

Besides capacities to innovate or imitate, the conditions of life could also influence how hard people work to create new ways of doing things and their willingness to adopt the latest invention. A common assumption is that "necessity is the mother of invention", that dire straits encourage people to consider new solutions to life's problems. While this position is reasonable one can easily envision how stressful conditions might promote conservatism as a psychological reaction, especially of the cost of making the wrong choice is very high. Some innovations in subsistence technology could in turn increase environmental carrying capacity, eventually leading to larger population sizes and higher densities, and creating new conditions of invention and social transmission. Some recent papers have explored potential feedback relationships between innovation rates and demographic factors.

Richerson et al. (2009) have undertaken the most explicit examination of innovation and demographic change, and the one most relevant to archaeology. They assume that significant innovations by definition lead to improvements in efficiency. They further assume that innovations occur probabilistically and that the frequency of invention correlates with conditions loosely described as population pressure (Richerson et al., 2009: 216–217). The model of Richerson and colleagues makes some interesting predictions about trajectories of growth and technological change. Among other things, the model predicts that for a range of "plausible" parameter settings, populations reach a period of prolonged steady growth, aided by periodic technological innovations, after about 1000 (model) years: this expectation contrasts sharply with the cycles of demographic growth and decline found in many other mammals. Although beneficial innovations may permit greater long-term population growth, they do not delay the onset of demographic stress in the model. Interestingly, stipulating that innovation is a probabilistic response to resource stress leads to the somewhat surprising prediction that evidence for population pressure will not necessarily precede significant improvements in technology (Richerson et al., 2009: 217). The authors then go on to discuss the implications of their model for large-scale cultural patterning during the late Pleistocene and early Holocene, including some of the historical phenomena described in this paper. Like purely demographic models, Richerson and colleagues' model better accounts for regional and temporal variation in evidence for "creativity" than do approaches based on essential properties of hominin species.

Efferson (2008) takes a somewhat different approach, exploring potential interactions between demography and economic productivity in humans, a species that can and does alter the productivity of its own resource base. He employs two related mathematical models, one from neoclassical economics and one from foraging ecology. Although he does not mention innovation rates specifically, Efferson permits consumers to increase economic productivity, which is equivalent to allowing only beneficial, adaptive innovations to proliferate. An analysis of the models predicts a complex series of outcomes regarding when and how consumer populations grow. Under some conditions, his analysis predicts burgeoning populations with constant per-capita resource budgets. In other situations, populations remain steady but per-capita resource budgets expand. This second scenario approximates the "demographic transition" experienced in many large-scale societies: whether it fits preindustrial situations remains to be seen. An important observation however is that the particular growth trajectory followed is strongly contingent on initial conditions, including what might be considered history or evolved propensities. This leads to a more general point. Although it makes good sense to explore the direct links between demography, network structure, innovation rates and eventually cognition, allowing these variables to influence one another creates nonlinear model systems. The models developed by Henrich, Shennan, Powell and others are essentially linear, characterised by unidirectional causality. Nonlinear models such as Efferson's may make for a more realistic approximation of the actual world, but inevitably they will also exhibit more complex and unpredictable dynamics as well as high levels of sensitivity to initial conditions.

6.2. CONCLUSION: SOME METHODOLOGICAL CHALLENGES

Creativity and innovation can be understood in different ways and studied at different scales. To a large extent, the way we think about and investigate innovation depends on the resolution implicit in our data. In the case of Palaeoanthopology, access to individual acts of creative genius is incomplete and fundamentally unpredictable. Palaeoanthropologists are in a much better position to evaluate what I have called emergent creativity, that is, places and periods characterised by very high levels of change or diversity. With reference to the title and theme of this volume, this paper makes the case that creativity and innovation can have their origins as much in demographic conditions and social dynamics as in individual cognitive processes. Moreover, as the case studies cited show, there is no single point of origin for emergent forms of creativity. Inherent, biologically fixed cognitive capacities must play a role, but variation at medium and small scales is likely to reflect fluctuating demographic conditions and changing strategies of network formation. As such, we should not be surprised to find evidence for multiple intervals of rapid change and heightened diversity interspersed with periods of stasis within a single biological or cultural lineage. Whether or not these phases of dynamism correspond to conventional psychological understandings of the creative process, such intervals are certainly important for investigating cultural evolutionary processes.

Research into the ways population sizes, population stability and social networks influence rates of culture change and scales of diversification is still in an early stage of development. Nonetheless, the initial results show that this is a productive way of thinking about long-term trends in hominin behavioural evolution. The approaches developed so far offer the potential to evaluate the fundamental question about cultural evolution and the nature of individual as well as aggregate creativity. For example, while it makes intuitive sense to assume that the invention and adoption of new behaviours are driven by hardship, by perceptions that existing options are inadequate, this is by no means axiomatic. There is heuristic value in imagining that innovation, like mutation, is essentially random. Moreover, the extent to which novel behaviours were randomly or deliberately produced may have changed over the course of human evolution, and this should have consequences for trajectories of change. If innovation is a truly stochastic phenomenon, then the frequency with which new behaviours appears should be a direct function of population sizes, and rates of innovation should closely track population growth, stability and decline. If, on the other hand, innovation is strongly influenced by per-capita resource availability, we can expect periods when innovation declined even as populations grew, particularly just after a new invention (or change in climate) raised the effective environmental carrying capacity. It would not be surprising to learn that there was a gradual shift from dynamics like those described by Powell et al. (2009) to the sort of regime imagined by Richerson et al. (2008) over the course of the Pleistocene.

Most of the studies reviewed in this paper are thought experiments and modelling exercises, establishing the plausibility of demography or network structures as determinants of the pace of culture change. Although demographically based approaches may provide quantitative predictions that are testable in principle, the field has not yet been able to provide many rigorous tests. Important obstacles in measurement must be solved before the models can be refined using empirical data. The two greatest challenges involve estimating population parameters and measuring rates of change and diversification in material culture.

Estimating absolute and relative population sizes in the Pleistocene is still a very inexact science (some even call it a "black art"). Direct methods for estimating population sizes or densities, such as counting sites or archaeological layers, counting radiometric dates or summing radiocarbon probability distributions, have their own advantages and technical weaknesses (see Chamberlain, 2006; Bocquet-Appel, 2008; Riede, 2008). Indirect approaches, such as using trends diet breadth (reviewed by Stiner and Munro, 2002; Munro and Atıcı, 2009), or other cultural responses as indicators of population densities involve extra inferential steps, and therefore also come with a risk of circularity when used naively as proxies for population levels. All these approaches at best provide relative measures of population census sizes. However, this should in no way discourage the further development of models based on demography. Progress in this area will inevitably be incremental: more refined models with more interesting predictions should lead to more rigorous tests and vice versa.

Measuring change and diversity in cultural expression presents equally difficult challenges, though of a different nature. All the examples of changes in rates of culture change or contrasts in diversity presented or cited here are nonquantitative. Because the cases discussed represent the observations of a great many researchers working independently, I am confident that they would bear up to more rigorous quantification. However, such qualitative assessments are insufficient to test models and refine their predictions. The most obvious solution, meta-analysis of published data, is hampered by the methods commonly used by prehistorians to describe artefacts. Studies of Palaeolithic material culture generally suffer from a low level of comparability. The exception is results from application of certain artefact typologies, but these are problematic for other reasons. Comparability is particularly an issue for research on processes of manufacture or *chaînes opératoires*, which provides the richest source of data on behavioural variation but normally presents it in the least standardised manner (Kuhn, 2010). Another obstacle is the tendency among archaeologists to describe what are presumed to be normative or modal forms rather than to assess variation in artefact forms and procedures (Bar-Yosef and Van Peer, 2009). Effectively documenting evolutionary processes requires accurate assessment of variation. In order to measure variability and change over time with precision, it is necessary to restudy many existing artefact collections using more specialised methodologies that enhance comparability. Some efforts have already been made in this direction (e.g., Tostevin, 2003; Hamilton and Buchanan, 2009; Lycett, 2009), though larger-scale studies are needed. Systematically reanalysing a great many previously published collections is a daunting prospect but the potential payoff could be enormous.

ACKNOWLEDGEMENTS

My thinking about the ideas and models discussed in this paper has benefitted greatly from the insights and knowledge of P. Jeffery Brantingham, W.R. Haas, J. Stephen Lansing, D. Shane Miller, Luke Premo, James Steele, and Mary Stiner. I am grateful to Scott Elias for organising this volume and asking me to participate.

REFERENCES

Agazi Negash, M., Shackley, M.S., 2006. Geochemical provenance of obsidian artifacts from the Middle Stone Age site of Porc Epic, Ethiopia. Archaeometry 48, 1–12.

Ambrose, S., 2001. Paleolithic technology and human evolution. Science 291, 1748–1753.

Amici, D., Aureli, F., Call, J., 2008. Fission-fusion dynamics, behavioral flexibility, and inhibitory control in primates. Curr. Biol. 18, 1415–1419.

Arrizabalaga, A., 2009. The Middle to Upper Paleolithic transition on the Basque crossroads: main sites, key issues. Mitt. Ges. Urgesch 18, 39–70.

Arzarello, M., Marcolini, F., Pavia, G., Pavia, M., Petronio, C., Petrucci, M., Rook, L., Sardella, R., 2007. Evidence of earliest human occurrence in Europe: the site of Pirro Nord (Southern Italy). Naturwissen 94, 107–112.

Bar-Yosef, O., 2002. The Upper Paleolithic revolution. Ann. Rev. Anthropol. 31, 363–393.

Bar-Yosef, O., 2006. Defining the Aurignacian. In: Bar-Yosef, O., Zilhão, Z. (Eds.), Towards a Definition of the Aurignacian. Trabalhos de Arqueologia, 45. Instituto Português de Arqueologia, Lisbon, pp. 11–18.

Bar-Yosef, O., Bordes, J.-G., 2010. Who were the makers of the Châtelperronian culture? J. Hum. Evol. 59, 586–593.

Bar-Yosef, O., Goren-Inbar, N., 1993. The Lithic Assemblages of Ubeidiya: A Lower Palaeolithic Site in the Jordan Valley. Institute of Archaeology, Hebrew University, Jerusalem.

Bar-Yosef, O., Van Peer, P., 2009. The chaîne opératoire method in Middle Paleolithic archeology. Curr. Anthropol. 50, 103–131.

Barkai, R., Lemorini, C., Shimelmitz, R., Lev, Z., Stiner, M.C., Gopher, A., 2009. A blade for all seasons? Making and using Amudian blades at Qesem Cave, Israel. Hum. Evol. 24, 57–75.

Basalla, G., 1988. The Evolution of Technology. Cambridge University Press, Cambridge.

Beaune, S., Coolidge, F., Wynn, T. (Eds.), 2009. Cognitive Archaeology and Human Evolution. Cambridge University Press, New York.

Beekman, M., Fathke, R.L., Seeley, T.D., 2006. How does an informed minority of scouts guide a honey bee swarm as it flies to its new home? Anim. Behav. 71, 161–171.

Bentley, R.A., 2009. Fashion versus reason in the creative industries. In: O'Brien, M.J., Shennan, S.J. (Eds.), Innovation in Cultural Systems: Contributions from Evolutionary Anthropology. MIT Press, Cambridge, MA, pp. 121–126.

Bentley, R.A., Hahn, M.W., Shennan, S.J., 2004. Random drift and culture change. Proc. R. Soc. Lond. B 271, 1443–1450.

Berna, F., Goldberg, P., 2008. Assessing Paleolithic pyrotechnology and associated hominin behavior in Israel. Israel J. Earth Sci. 56, 107–121.

Bettencourt, L.M.A., Lobo, J., Helbing, D., Kühnert, C., West, G.B., 2007. Growth, innovation, scaling, and the pace of life in cities. Proc. Natl. Acad. Sci. U S A 104, 7301–7306.

Binford, L.R., 2001. Constructing Frames of Reference: An Analytical Method for Archaeological Theory Building Using Hunter–Gatherer and Environmental Data Sets. University of California Press, Berkeley.

Bocquet-Appel, J.P. (Ed.), 2008. Recent Advances in Paleodemography: Data, Techniques, Patterns. Springer, Amsterdam.

Bocquet-Appel, J.-P., Tuffreau, A., 2009. Technological responses of Neanderthals to macroclimatic variations (240,000–40,000 BP). Hum. Biol. 81, 287–307.

Boëda, E., 1991. Approche de la variabilite des systemes de production lithique des industries du Paleolithique inferieur et moyen: chronique d'une variabilite attendue. Techniques et Cult 17-18, 37–79. in French.

Boëda, E., Hou, Y.-M., 2010. Analyse des artefacts lithiques du site de Longgupo. L'Anthropol 115, 78–175. in French w/English abstract.

Boëda, E., Geneste, J.M., Meignen, L., 1990. Identification de chaînes operatoires lithiques du Paleolithique ancien et moyen. Paleo 2, 43–80. in French w/English abstract.

Bolus, M., Conard, N., 2001. The late Middle Paleolithic and earliest Upper Paleolithic in Central Europe and their relevance for the Out of Africa hypothesis. Quaternary Int. 75, 29–40.

Bon, F., 2006. A brief overview of Aurignacian cultures in the context of the industries of the transition from the Middle to the Upper Paleolithic. In: Bar-Yosef, O., Zilhão, Z. (Eds.), Towards a Definition of the Aurignacian. Trabalhos de Arqueologia, 45. Instituto Português de Arqueologia, Lisbon, pp. 133–144.

Boone, J.L., 2002. Subsistence strategies and early human population history: an evolutionary ecological perspective. World Archaeol. 34, 6–25.

Bordes, F., 1961. Typologie du Paléolithique Ancien et Moyen. Delmas, Bordeaux.

Bordes, J.G., 2006. News from the West: are-evaluation of the classic Aurignacian sequence from the Pèrigord. In: Bar-Yosef, O., Zilhão, Z. (Eds.), Towards a Definition of the Aurignacian. Trabalhos de Arqueologia, 45. Instituto Português de Arqueologia, Lisbon, pp. 147–151.

Bordes, J.-G., Bon, F., Le Brun-Ricalens, F., 2005. Le transport des matières premières lithiques à l'Aurignacienancien entre le Nord et Sud de l'Aquitaine: faits attendus, faits nouveaux. In: Jaubert, J., Barbaza, M. (Eds.), 2005. Déplacements, Mobilité, Echanges Durant la Préhistoire, 126e. Congrès National des Sociétés Historiques et Scientifiques, Toulouse, pp. 185–198.

Borgatti, S.P., Mehra, A., Brass, D.J., Labianca, G., 2009. Network analysis in the social sciences. Science 323, 892–895.

Boyd, R., Richerson, P., 1985. Culture and the Evolutionary Process. University of Chicago Press, Chicago.

Boyd, R., Richerson, P., 2005. The Origin and Evolution of Cultures. Oxford University Press, Oxford.

Brain, C.K., Sillen, A., 1988. Evidence from the Swartkrans cave for the earliest use of fire. Nature 336, 464–466.

Bramble, D.M., Lieberman, D.E., 2004. Endurance running and the evolution of Homo. Nature 432, 345–352.

Brantingham, P.J., 2003. A neutral model of stone raw material procurement. Am. Antiq 68, 487–509.

Brantingham, P.J., 2006. Measuring forager mobility. Curr. Anthropol. 47, 435–459.

Brown, K.S., Marean, C.W., Herries, A.I.R., Jacobs, Z., Tribolo, C., Braun, D., Roberts, D.L., Meyer, M.C., Bernatchez, J., 2009. Fire as an engineering tool of early modern humans. Science 325, 859–862.

Burch, E.S. Jr., 1991. Modes of exchange in north-west Alaska. In: Ingold, T., Riches, D., Woodburn, J. (Eds.), Hunters and Gatherers. Property, Power, and Ideology, vol. 2. Berg, Oxford, pp. 95–109.

Carbonell, E., García-Antón, M.D., Mallol, C., Mosquera, M., Ollé, A., Rodríguez, X.P., Sahnouni, M., Sala, R., Vergès, J.M., 1999. The TD6 level lithic industry from Gran Dolina, Atapuerca (Burgos, Spain): production and use. J. Hum. Evol. 37, 653–693.

Carruthers, P., Chamberain, A. (Eds.), 2002. Evolution and the Human Mind: Modularity, Language and Meta-Cognition. Cambridge University Press, Cambridge.

Casanova I Martí, J., Martínez Moreno, J., Mora Turcal, R., de la Torre, I., 2009. Stratégies techniques dans le Paléolithique moyen du sud-est des Pyrénées. L'Anthropol. 113, 313–340.

Centola, D., Macy, M., 2007. Contagions and the weakness of long ties. Am. J. Sociol. 113, 702–734.

Chabai, V., 2003. The chronological and industrial variability of the Middle to Upper Paleolithic transition in eastern Europe. In: Bar-Yosef, O., Zilhão, Z. (Eds.), Towards a Definition of the Aurignacian. Trabalhos de Arqueologia, 45. Instituto Português de Arqueologia, Lisbon, pp. 71–86.

Chamberlain, A.T., 2006. Demography in Archaeology. Cambridge University Press, Cambridge.

Clottes, J., 2003. Chauvet Cave: The Art of Earliest Times. Paul G. Bahn (translator). University of Utah Press, Salt Lake City.

Collard, M., Kemery, M., Banks, S., 2005. Causes of tool kit variation among hunter-gatherers: a test of four competing hypotheses. Can. J. Archaeol. 29, 1–19.

Conard, N.J., 2009. A female figurine from the basal Aurignacian of Hohle Fels Cave in southwestern Germany. Nature 459, 248–252.

Conard, N.J., Malina, M., Münzel, S.C., 2009. New flutes document the earliest musical tradition in southwestern Germany. Nature 460, 737–740.

Coolidge, F., Wynn, T., 2009. The Rise of *Homo sapiens*: The Evolution of Modern Thinking. Wiley Blackwell, West Sussex.

Copeland, L., 2000. Yabrudian and related industries: the state of research in 1996. In: Ronen, A., Weinstein-Evron, M. (Eds.), BAR International Series 850, Oxford, pp. 97–117.

Cowan, R., Jonard, N., 2004. Network structure and the diffusion of knowledge. J. Econ. Dynam. Control 28, 1557–1575.

de la Torre, I., 2011. The origins of stone tool technology in Africa: a historical perspective. Phil. Trans. R. Soc. B 366, 1028–1037.

de Lumley, H., Barrier, P., Cauche, D., Grégoire, S., 2010. Les Industries Lithiques Archaïques de Barranco Leon et de Fuente Nueva 3: Orce, Bassin de Guadix-Baza, Andalousie. CNRS, Paris. in French w/English abstracts.

Deacon, H.J., 1989. Late Pleistocene palaeoecology and archaeology in the southern Cape, South Africa. In: Mellars, P., Stringer, C. (Eds.), The Human Revolution: Behavioural and Biological Perspectives on the Origins of Modern Humans. Edinburgh University Press, Edinburgh, pp. 4547–4564.

Delagnes, A., Meignen, L., 2005. Diversity of lithic production systems in the Middle Paleolithic in France. Are there any chronological trends. In: Hovers, E., Kuhn, S. (Eds.), Transitions before the Transition: Evolution and Stability in the Middle Paleolithic and Middle Stone Age. Springer, New York, pp. 85–107.

Dennel, R., 2009. The Palaeolithic Settlement of Asia. Cambridge University Press, Cambridge.

Djindjian, F., Kozlowski, J., Basile, F., 2003. Europe during the early Upper Paleolithic: a synthesis. In: Zilhão, J., d'Errico, F. (Eds.), The Chronology of the Aurignacian and of the Transitional Technocomplexes. Dating, Stratigraphies, Cultural Implications. Trabalhos de Arqueologia, 33. Instituto Português de Arqueologia, Lisbon, pp. 29–47.

Duke, C., Steele, J., 2009. Geology and lithic procurement in Upper Palaeolithic Europe: a Weights-of-Evidence based GIS model of lithic resource potential. J. Archaeol. Sci. 37, 813–824.

Dunbar, R.I.M., 1996. Groming, Gossip, and the Evolution of Language. Harvard University Press, Cambridge, MA.

Dunbar, R.I.M., 1998. The social brain hypothesis. Evol. Anthropol. 6, 178–190.

Dunbar, R.I.M., Shultz, S., 2007. Evolution in the social brain. Science 317, 1344–1347.

D'Errico, F., Nowell, A., 2000. A new look at the Berekhat Ram figurine: implications for the origins of symbolism. Camb. Archaeol. J. 10, 123–167. with commentary.

D'Errico, F., Stringer, C., 2011. Evolution, revolution or saltation scenario for the emergence of modern cultures? Phil. Trans. R. Soc. B 366, 1060–1069.

Efferson, C., 2008. Prey-producing predators: the ecology of human intensification. Nonlinear Dynamics Psychol. Life Sci. 12, 55–74.

Faisal, A., Stout, D., Apel, J., Bradley, B., 2010. The manipulative complexity of lower Paleolithic stone toolmaking. PLoS One 5, e13718. http://dx.doi.org/10.1371/journal.pone.0013718.

Feblot-Augustins, J., 1993. Mobility strategies in the Late Middle Palaeolithic of Central Europe and Western Europe: elements of stability and variability. J. Anthropol. Archaeol. 12, 211–265.

Feblot-Augustins, J., 1997. La Circulation des Matières Premières au Paléolithique, vol. 1. et 2. ERAUL 75, Liège. (in French).

Feblot-Augustins, J., 2009. Revisiting European Upper Palaeolithic raw material transfers: the demise of the cultural ecological paradigm? In: Adams, B., Blades, B.S. (Eds.), Lithic Materials and Paleolithic Societies. Wiley Blackwell, New York, pp. 25–46.

Fedele, F.G., Giaccio, B., Hajdas, I., 2008. Time scales and cultural process at 40,000 BP in the light of the Campanian Ignimbrite eruption, western Eurasia. J. Hum. Evol. 55, 834–857.

Foley, R., Lahr, M.-M., 2003. On stony ground: lithic technology, human evolution and the emergence of culture. Evol. Anthropol. 12, 109–122.

Gamble, C., 1996. Timewalkers. The Prehistory of Global Colonization. Harvard University Press, Cambridge, MA.

Gamble, C., 1997. Paleolithic society and the release from proximity: a network approach to intimate relations. World Archaeol. 29, 426–449.

Gamble, C., 1999. Paleolithic Societies of Europe. Cambridge University Press, Cambridge.

Golovanona, L.V., Doronichev, V.B., Cleghorn, N.E., Koulkova, M.A., Sapelko, T.V., Shackley, M.S., 2010. Significance of ecological factors in the Middle to Upper Paleolithic transition. Curr. Anthropol. 51, 655–692.

Goren-Inbar, N., Alperson, N., Kislev, M.E., Simchoni, O., Melamed, Y., Ben-Nun, A., Werker, E., 2004. Evidence of hominin control of fire at Gesher Benot Ya'aqov, Israel. Science 304, 725–727.

Granovetter, M., 1973. The strength of weak ties. Am. J. Sociol. 78, 1360–1380.

Grove, M., 2009. Hunter–gatherer movement patterns: causes and constraints. J. Anthropol. Archaeol. 28, 2–233.

Güleç, E., White, T., Howell, F.C., Özer, I., Sağır, M., Yılmaz, H., Kuhn, S., 2009. The lower Pleistocene lithic assemblage from Dursunlu (Konya), central Anatolia, Turkey. Antiquity 83, 11–22.

Hahn, J., 1972. Aurignacian signs, pendants and art objects in central and eastern Europe. World Archaeol. 3, 252–266.

Hahn, M., Bentley, A., 2003. Drift as a mechanism for cultural change: an example from baby names. Proc. R. Soc. Lond. B 270 (Supplement) S120–S12.

Hamilton, M.J., Buchanan, B., 2009. The accumulation of stochastic copying errors causes drift in culturally transmitted technologies:

quantifying Clovis evolutionary dynamics. J. Anthropol. Archaeol. 28, 55–69.

Hamilton, M.J., Milne, B.T., Walker, R.S., Brown, J.H., 2007. Nonlinear scaling of space use in human hunter–gatherers. Proc. Natl. Acad. Sci. U S A 104, 4765–4769.

Hawks, J., Wang, E., Cochran, G.M., Harpending, H., Moyzis, R.K., 2007. Recent acceleration of human adaptive evolution. Proc. Natl. Acad. Sci. U S A 104, 20753–20758.

Heinrich, H., 1988. Origin and consequences of cyclic ice rafting in the Northeast Atlantic Ocean during the past 130,000 years. Quaternary Res. 29, 142–152.

Henrich, J., 2004. Demography and cultural evolution: how adaptive cultural processes can produce maladaptive losses: the Tasmanian case. Am. Antiq. 69, 197–214.

Henrich, J., Boyd, R., Richerson, P., 2008. Five misunderstandings about cultural evolution. Hum. Nat. 19, 119–137.

Hoffecker, J., 2005. Innovation and technological knowledge in the Upper Paleolithic of northern Eurasia. Evol. Anthropol. 14, 186–198.

Hoffecker, J., 2011. The early upper Paleolithic of eastern Europe reconsidered. Evol. Anthropol. 20, 24–39.

Hovers, E., Kuhn, S. (Eds.), 2005. Transitions before the Transition: Evolution and Stability in the Middle Paleolithic and Middle Stone Age. Springer, New York.

Hubbell, S.P., 2001. The Unified Neutral Theory of Biodiversity and Biogeography. Princeton University Press, Princeton, NJ.

Jacobs, Z., Roberts, R.G., 2008. Testing times: old and new chronologies for the Howiesons Poort and Still Bay Industries in environmental context. S. Afric. Archaeol. Soc. Goodwin Ser. 10, 9–34.

Jacobs, Z., Roberts, R.G., Galbraith, R.F., Deacon, H.J., Grün, R., Mackay, A., Mitchell, P., Vogelsang, R., Wadley, L., 2008. Ages for the Middle Stone Age of Southern Africa: implications for human behavior and dispersal. Science 322, 733–735.

Jelinek, A.J., 1990. The Amudian in the context of the Mugharan tradition at the Tabun Cave (Mount Carmel), Israel. In: Mellars, P. (Ed.), The Emergence of Modern Humans. Cornell University Press, Ithica, NY, pp. 81–90.

Karkanas, P., Koumouzelis, M., Kozlowski, J.K., Sitlivy, V., Sobczyk, K., Berna, F., Wiener, S., 2004. The earliest evidence for clay hearths: Aurignacian features in Klisoura Cave southern Greece. Antiquity 78, 513–525.

Karkanas, P., Shahack-Gross, R., Ayalon, A., Bar-Matthews, M., Barkai, R., Fromkin, A., Gopher, A., Stiner, M.C., 2007. Evidence of habitual use of fire at the end of the lower Paleolithic: site-formation processes at Qesem Cave, Israel. J. Hum. Evol. 53, 197–212.

Kelly, R.L., 1983. Hunter–gatherer mobility strategies. J. Anthropol. Res. 39, 277–306.

Kelly, R.L., 1995. The Foraging Spectrum: Diversity in Hunter–Gatherer Lifeways. Smithsonian Institution Press, Washington.

Klein, R., 2000. Archaeology and the evolution of human behavior. Evol. Anthropol. 9, 17–36.

Klein, R., Edgar, B., 2002. The Dawn of Human Culture. Wiley, New York.

Kline, M.A., Boyd, R., 2010. Population size predicts technological complexity in Oceania. Proc. R. Soc. Lond. B 277, 2559–2564.

Kuhn, S., 1992. On planning and curated technologies in the Middle Paleolithic. J. Anthropol. Res. 48, 185–214.

Kuhn, S., 2005. Trajectories of change in the Middle Paleolithic of Italy. In: Hovers, E., Kuhn, S. (Eds.), Transitions before the Transition: Evolution and Stability Change in the Middle Paleolithic and Middle Stone Age. Springer, New York, pp. 109–120.

Kuhn, S., 2010. On standardization in the Paleolithic: measures, causes, and interpretations of metric similarity in stone tools. In: Nowell, A., Davidson, I. (Eds.), Lithic Technology and the Evolution of Human Cognition. University Press of Colorado, Boulder, pp. 105–134.

Kuhn, S., 2011. Neanderthal technoeconomics. An assessment and suggestions for future developments. In: Richeter, J., Conard, N. (Eds.), Neanderthal Lifeways, Subsistence and Technology. Springer, New York, pp. 99–100.

Kuhn, S., Stiner, M.C., 1998. Middle Paleolithic creativity: reflections on an oxymoron? In: Mithen, S. (Ed.), Creativity in Human Evolution and Prehistory. Routledge, London, pp. 146–164.

Laland, K., 2004. Social learning strategies. Learn. Behav. 32, 4–14.

Lemonnier, P., 1992. Elements for an Anthropology of Technology. In: Anthropological Papers, Museum of Anthropology, vol. 88. University of Michigan Press, Ann Arbor.

Lombard, M., Parsons, I., 2010. Fact or fiction? Behavioural and technological reversal after 60 ka in southern Africa. S. Afric. Archaeol. Bull. 65, 221–228.

Lycett, S.J., 2009. Understanding ancient hominin dispersals using artefactual data: a phylogeographic analysis of Acheulean handaxes. PLoS One 4 (10), e7404. 1–6.

Lycett, S.J., Norton, C., 2010. A demographic model for Palaeolithic technological evolution: the case of East Asia and the Movius Line. Quaternary Int. 211, 55–65.

Maslin, M., Seidov, D., Lowe, J., 2001. Synthesis of the nature and causes of rapid climate transitions during the Quaternary. Geophys. Monog. 126, 9–52.

Mazza, P.P.A., Martini, F., Sala, B., Magi, B., Columbini, M.P., Giachi, G., Landucci, F., Lemorini, C., Modugno, F., Ribechini, E., 2006. A new Palaeolithic discovery: tar-hafted stone tools in a European Mid-Pleistocene bone-bearing bed. J. Archaeol. Sci. 33, 1310–1318.

McBrearty, S., Brooks, A., 2000. The revolution that wasn't: a new interpretation of the origin of modern human behavior. J. Hum. Evol. 39, 453–563.

McCall, G.S., 2007. Behavioral ecological models of lithic technological change during the later Middle Stone Age of South Africa. J. Archaeol. Sci. 34, 1738–1751.

McElreath, R., 2010. The coevolution of genes, innovation and culture in human evolution. In: Silk, J., Kappeler, P. (Eds.), Mind the Gap: Tracing the Origins of Human Universals. Springer, New York, pp. 451–474.

McPherron, S., Alemseged, Z., Marean, C., Wynn, J., Reed, D., Geraads, D., Bobe, R., Béarat, H., 2010. Evidence for stone-tool-assisted consumption of animal tissues before 3.39 million years ago at Dikika, Ethiopia. Nature 466, 857–860.

Meignen, L., 2007. Le phénomène laminaire au Proche-Orient, du Paléolithique Inférieur aux débuts du Paléolithique Supérieur. In: Evin, J. (Ed.), 2007. Congrès du Centenaire: Un Siècle de Construction du Discours Scientifique en Préhistoire, XXVI. Congrès Préhistorique de France, pp. 79–94. Avignon 2004. Paris.

Meignen, L., Bar-Yosef, O., Goldberg, P., 2001. Le feu au Paléolithique moyen: recherches sur les structures de combustion et le statut des foyers. L'exemple du Procbe-Orient. Paleorient 26, 9–22.

Meignen, L., Delagnes, A., Bourguignon, L., 2009. Patterns of lithic raw material procurement and transformation during the Middle

Paleolithic in western Europe. In: Adams, B., Blades, B. (Eds.), Lithic Materials and Paleolithic Societies. Wiley Blackwell, New York, pp. 15–24.

Mellars, P., 2006. Archeology and the dispersal of modern humans in Europe: deconstructing the Aurignacian. Evol. Anthropol. 15, 167–182.

Merrick, H.V., Brown, F.H., Nash, W., 1994. Use and movement of obsidian in the Early and Middle Stone Ages of Kenya and Northern Tanzania. In: Childs, S.T. (Ed.), 1994. Society, Culture and Technology in Africa. MASCA Research Papers in Science and Archaeology, Vol. 11. (Supplement), pp. 29–44. Philadelphia.

Meyers, M.A., 2007. Happy Accidents: Serendipity in Modern Medical Breakthroughs. Arcade, New York.

Mithen, S., 1996. The Prehistory of the Mind. Thames and Hudson, London.

Movius, H., 1948. The lower Paleolithic cultures of southern and eastern Asia. Trans. Amer. Philos. Soc. 38, 329–420.

Munro, N., Atıcı, L., 2009. Human subsistence change in the late Pleistocene Mediterranean: the status of research on faunal intensification, diversification and specialization. Before Farming 2009/3 (online version), article 1.

Newman, M., Barabasi, A.L., Watts, D.J. (Eds.), 2006. The Structure and Dynamics of Networks. Princeton University Press, Princeton, NJ.

Noble, W., Davidson, I., 1996. Human Evolution, Language and Mind. A Psychological and Archaeological Enquiry. Cambridge University Press, New York.

Parfitt, S.A., Barendregt, R.W., Breda, M., Candy, I., Collins, M.J., Russell Coope, G., Durbridge, P., Field, M.H., Lee, J.R., Lister, A.M., Mutch, R., Penkman, K.E.H., Preece, R.C., Rose, J., Stringer, C.B., Symmons, R., Whittaker, J.E., Wymer, J.J., Stuart, A.J., 2005. The earliest record of human activity in Northern Europe. Nature 438, 1008–1012.

Peck, J.R., Barreau, G., Heath, S.C., 1997. Imperfect genes, Fisherian mutation and the evolution of sex. Genetics 145, 1171–1199.

Pelegrin, J., 1990. Prehistoric lithic technology: some aspects of research. Archaeol. Rev. Camb 9, 116–125.

Perreault, C., Brantingham, P.J., 2010. Mobility-driven cultural transmission along the forager–collector continuum. J. Anthropol. Archaeol. 30, 62–68.

Petraglia, M.D., Clarkson, C., Boivin, N., Haslam, M., Korisettar, R., Chaubey, G., Ditchfield, P., Fuller, D., James, H., Jones, S., Kivisild, T., Koshy, J., Lahr, M.M., Metspalu, M., Roberts, R., Arnold, L., 2009. Population increase and environmental deterioration correspond with microlithic innovations in South Asia ca. 35,000 years ago. Proc. Natl. Acad. Sci. U S A 106, 12261–12266.

Plummer, T.L., 2004. Flaked stones and old bones: biological and cultural evolution at the dawn of technology. Year b. Phys. Anthropol. 47, 118–164.

Powell, A., Shennan, S., Thomas, M., 2009. Late Pleistocene demography and the appearance of modern human behavior. Science 324, 1298–1301.

Premo, L., 2012. Demographic resilience, connectedness, and cultural evolution in structured populations. Advances in Complex Systems 15, DOI No: 10.1142/S0219525911003268.

Premo, L., Kuhn, S.L., 2010. Modeling effects of local extinctions on culture change and diversity in the Paleolithic. PLoS One 5 (12), e15582.

Read, D., 2006. Tasmanian knowledge and skill: maladaptive imitation or adequate technology. Am. Antiq. 71, 164–184.

Read, D., 2008. An interaction model for resource implement complexity based on risk and number of annual moves. Am. Antiq. 73, 599–625.

Reader, S., Laland, K., 2002. Social intelligence, innovation, and enhanced brain size in primates. Proc. Natl. Acad. Sci. U S A 99, 4436–4441.

Richerson, P., Boyd, R., Bettinger, R., 2009. Cultural innovations and demographic change. Hum. Biol. 81, 211–235.

Riede, F., 2008. Climate and demography in early prehistory: using calibrated ^{14}C Dates as population proxies. Hum. Biol. 81, 309–337.

Rigaud, J.-P., Simek, J.F., Ge, T., 1995. Mousterian fires from Grotte XVI (Dordogne, France). Antiquity 69, 902–912.

Roebroeks, W., 2006. The human colonisation of Europe: where are we? J. Quaternary Sci. 21, 425–435.

Rogers, E.M., 1983. Diffusion of Innovations. Free Press, New York.

Santoja, M., Villa, P., 2006. The Acheulean of western Europe. In: Goren-Inbar, N., Sharon, G. (Eds.), Axe Age, Acheulian Toolmaking from Quarry to Discard. Equinox, London, pp. 429–478.

Schiffer, M.B., 1992. Technological Perspectives on Behavioral Change. University of Arizona Press, Tucson.

Schiffer, M.B., 2002. Studying technological differentiation: the case of 18th-century electrical technology. Am. Anthropol. 104, 1148–1161.

Schweizer, T., 1996. Reconsidering social networks: reciprocal gift exchange among the !Kung. J. Quantit. Anthropol. 6, 147–170.

Seeley, T.D., Visscher, P.K., Passino, K.M., 2006. Group decision making in honey-bee swarms. Am. Sci. 94, 220–229.

Shennan, S., 2001. Demography and cultural innovation: a model and its implications for the emergence of modern human culture. Camb. Archaeol. J. 11, 5–16.

Shennan, S., Edinborough, K., 2007. Prehistoric population history: from the late Glacial to the Late Neolithic in Central and Northern Europe. J. Archaeol. Sci. 34, 1339–1345.

Slimak, L., 2008. The Neronian and the historical structure of cultural shifts from Middle to Upper Palaeolithic in Mediterranean France. J. Archaeol. Sci. 35, 2204–2214.

Slimak, L., Giraud, Y., 2007. Circulations sur plusieurs centaines de kilomètres durant le Paléolithique moyen. Contribution à la connaissance des sociétés néandertaliennes. Comptes Rendus Palevol 6, 359–368. in French/English abstract.

Soressi, M., 2006. Late Mousterian lithic technology: its implications for the pace of the emergence of behavioural modernity and the relationship between behavioural modernity and biological modernity. In: Blackwell, L., d'Errico, F. (Eds.), From Tools to Symbols: From Early Hominids to Modern Humans. Witwatersrand University Press, Witwatersrand, pp. 389–417.

Stiner, M., Kuhn, S., 2006. Changes in the 'connectedness' and resilience of Paleolithic societies in Mediterranean ecosystems. Hum. Ecol. 34, 693–712.

Stiner, M., Munro, N., 2002. Approaches to prehistoric diet breadth, demography and prey ranking systems in time and space. J. Archeol. Meth. Theory 9, 181–214.

Stout, D., 2011. Stone toolmaking and the evolution of human culture and cognition. Phil. Trans. R. Soc. B 366 1060–1059.

Straus, L., 2010. The emergence of modern-like forager capacities & behaviors in Africa and Europe: abrupt or gradual, biological or demographic? Quaternary Int. http://dx.doi.org/10.1016/j.quaint.2010.10.002 Key: citeulike:8066065.

Taborin, Y., 1993. Le Parure en Coquillage au Paléolithique, XXIXe. Supplément à Gallia Préhistoire, Paris (in French).

Texier, P.J., Porraz, G., Parkington, J., Rigaud, J.P., Poggenpoel, C., Miller, C., Tribolo, C., Cartwright, C., Coudenneau, A., Klein, R., Steele, T., Verna, C., 2010. A Howiesons Poort tradition of engraving ostrich eggshell containers dated to 60,000 years ago at Diepkloof Rock Shelter, South Africa. Proc. Natl. Acad. Sci. U S A 107, 6180–6185.

Teyssandier, N., 2007. L'émergence du Paléolithique supérieur en Europe: mutations culturelles et rythmes d'évolution. Paléo 19, 367–389. in French w/English abstract.

Tostevin, G.B., 2003. Attribute analysis of the lithic technologies of Stránská skála II-III in their regional and inter-regional context. In: Svoboda, J., Bar-Yosef, O. (Eds.), Sránská skála: Origins of the Upper Paleolithic in the Brno Basin. Peabody Museum, Harvard University, Camridge, MA, pp. 77–118.

Vandiver, P.B., Sofer, O., Klima, B., Svoboda, J., 1989. The origin of ceramic technology at Dolni Vestonice, Chechoslovakia. Science 246, 1002–1008.

Villa, P., 2009. The Lower to Middle Paleolithic transition. In: Camps, M., Chuahan, R. (Eds.), Sourcebook of Paleolithic Transitions. Springer, New York, pp. 265–270.

Vishnyatsky, L.B., 2000. The Pre-Aurignacian and Amudian as intra-Yabrudian episode. In: Ronen, A., Weinstein-Evron, M. (Eds.), Toward Modern Humans: Yabrudian and Micoquian, 400-50 k years Ago. BAR International Series, vol. 850, pp. 145–151. Oxford.

Watts, D.J., Strogatz, S.H., 1998. Collective dynamics of 'small-world' networks. Nature 393, 440–442.

Whallon, R., 2006. Social networks and information: non-"utilitarian" mobility among hunter–gatherers. J. Anthropol. Archaeol. 25, 259–270.

Wiessner, P., 1977. Hxaro: a regional system of reciprocity for reducing risk among the !Kung San. University Microfilms, Ann Arbor.

Wiessner, P., 1982. Risk, reciprocity and social influences on !Kung San economics. In: Leacock, E., Lee, R.B. (Eds.), Politics and History in Band Societies. Cambridge University Press, Cambridge, pp. 61–84.

Wrangham, W., 2009. Catching Fire: How Cooking Made us Human. Basic Books, New York.

Wynn, T., Coolidge, F.L., 2004. The expert Neandertal mind. J. Hum. Evol. 46, 467–487.

Yengoyan, A., 1979. Economy, society, and myth in aboriginal Australia. Ann. Rev. Anthropol. 8, 393–415.

Zilhào, Z., d'Errico, F., 2003. The chronology of the Aurignacian and Transitional technocomplexes: where do we stand? In: Zilhão, J., d'Errico, F. (Eds.), The Chronology of the Aurignacian and of the Transitional Technocomplexes. Dating, Stratigraphies, Cultural Implications. Trabalhos de Arqueologia, 33. Instituto Português de Arqueologia, Lisbon, pp. 334–349.

Chapter 7

The Evolutionary Ecology of Creativity

John F. Hoffecker

Institute of Arctic and Alpine Research, University of Colorado, Boulder, CO 80309 USA

"Creation is not fashioning something out of nothing, but refashioning what already is."

V. Gordon Childe, *Society and Knowledge* (1956: 124)

Creativity is the recombination of informational units into novel arrangements or structures. In a sense, the evolutionary process is "creative", because it entails recombination of informational units in the form of DNA sequences into a potentially infinite variety of arrangements, and these, in turn, may be transformed into organisms that exhibit evolutionary "innovations" such as colour vision or feathers (e.g., Nitecki, 1990). Humans are creative with informational units in the form of synaptic connections in the brain and they have the capacity for transforming these units into other forms of information, such as a poem, and into structures based on information, such as an armchair (e.g., Dawkins, 1976: 203–215).

An emphasis on the "forming of associative elements into new combinations" is evident among the many psychologists (e.g., Mednick, 1962: 220–232) who addressed the issue of creativity during the mid-twentieth century (Kyriacou, 2009: 15–24). But in recent decades, it also became a focal point among linguists, attempting to explain what Noam Chomsky (2002: 55) referred to as "the ordinary creative use of language". The ability "to generate an infinite range of expressions from a finite set of elements" (or *discrete infinity*) has been termed the "core property" of syntactic language (Chomsky, 1988: 169–170; Hauser et al., 2002). The same property may be applied to virtually every sphere of human activity – from cooking recipes to organization of domestic space, from dance movements to polychrome paintings, and from clothing design to a piano concerto (Corballis, 2003; Hoffecker, 2007).

An essential characteristic of the structures created by the evolutionary process and the human imagination is *hierarchical* organization. There are only four DNA base pairs, and they are arranged in groups of three ("codons"), each of which codes for one of twenty amino acids. But varying sequences of amino acids form different proteins, and these in turn are building blocks for more complex structures. Even the comparatively simple prokaryote is based on a highly complex, hierarchically-organized design, and the potential variety of multi-cellular plant and animal life seems infinite. By the same token, human language is "organized like the genetic code – hierarchical, generative, recursive, and virtually limitless with respect to scope of expression" (e.g., Hauser et al., 2002: 1569).[1] Discrete sounds produced by the vocal tract are combined into larger units (words), which are in turn combined into larger units (phrases), which are then combined into larger units (sentences), and so forth. Again, the same principle applies to other spheres of activity, such as making tools or composing music.

How did humans acquire their unique powers to creatively combine and recombine informational units in the brain? To begin with, humans gather, store, and share an enormous quantity of non-genetic information. Humans collect so much information that they evolved grotesquely over-sized crania (relative to their body size) to store it (e.g., McHenry, 1994; Klein, 2009), and later had to devise a variety of technologies to store increasingly large amounts of it outside the brain (e.g., Donald, 1991: 269–360; Renfrew, 1998). With an estimated 10 billion neurons, each of which is connected to 1000 or more other neurons (Fine, 2008: 27–33), the modern human brain has been described as "the most complex material object in the known universe" (Edelman, 2004: 14–19).

Secondly, despite the fact that modern humans seem to collect an extraordinary amount of useless information (also true of a genome) it is apparent that our social and economic life demand substantial data gathering and processing. Even in small social settings, each person collects an immense quantity of highly detailed information pertaining to relatives, friends, colleagues, enemies, acquaintances, and

---

1. Experimental research with tamarins, which can be taught a "finite state grammar" that governs arrangements of a small set of elements, but cannot learn "phrase structure grammar", illustrates the limitations of a non-hierarchically-organized information system (Fitch and Hauser, 2004).

others. Robin Dunbar (1996) has argued that the major increases in later *Homo* brain volume are tied to social networking. At the same time, even the simplest foraging economies are heavily based on the gathering, processing, and sharing of information (e.g., Mithen, 1990: 52–88).

In sum, not only are humans awash – at this point virtually drowning – in the informational units that they have learned to manipulate so creatively (see Gleick, 2011), but they employ them daily to achieve ends that have significant consequences for their survival and long-term reproductive success.

The key to understanding how humans became creative lies in their *translation* of information stored in the neuronal networks of the brain to other entities. At some point in their evolution, humans started translating sets of informational units in their brains to other forms outside the brain. They probably began with complex patterned arrangements that were devoid of information as courtship displays, which is the simplest and most common form among non-human taxa. Early humans also probably evolved a faculty for translating neuronal information to another form of information as vocalizations and/or gestures, which is much rarer among living animals. But the critical development seems to have been a unique human faculty for translating information to complex, hierarchically-organized artefacts. As a consequence of this development, humans entered into a dynamic relationship with the informational units of the brain. They began to manipulate their thoughts in translated form outside the brain in a setting exposed to both social and environmental feedback. It was from this dynamic setting that the creative recombination of informational units in the brain emerged with profound effects on both the creators and their world.

7.1. MODERN HUMANS AS A 'MAJOR TRANSITION' IN EVOLUTION

"Our basic idea was that evolution depends on changes in the information that is passed between generations, and that there have been 'major transitions' in the way that information is stored and transmitted . . ."
John Maynard Smith and Eörs Szathmáry, *The Origins of Life* (1999: vii)

During the final years of his highly productive career, John Maynard Smith co-authored two books with Eörs Szathmáry on the *major transitions* in evolution. Maynard Smith and Szathmáry (1999: 17, Table 2.2) identified eight such transitions: (1) replicating molecules to populations of molecules in proto-cells; (2) independent replicators to chromosomes; (3) RNA as gene and enzyme to DNA genes and protein enzymes; (4) bacterial cells (prokaryotes) to cells with nuclei and organelles (eukaryotes); (5) asexual clones to sexual populations; (6) single-celled organisms to animals, plants, and fungi; (7) solitary individuals to colonies with non-reproductive castes (eusocial animals or "super-organisms"); and (8) primate societies to human societies and language (see also Maynard Smith and Szathmáry, 1995: 6, Table 1.2).

Each transition is completed when "entities that were capable of independent replication before the transition could afterwards replicate only as part of a larger whole" (Maynard Smith and Szathmáry, 1999: 19). Eukaryotes or true cells, for example, cannot replicate as a collection of independent organelles, and an ant colony can replicate only as a unit (i.e., through the queen). All of the major transitions in evolution are characterized by "changes in the way in which information is stored, transmitted, and translated" (Maynard Smith and Szathmáry, 1999: 16).

By including human language in their list of major transitions, Maynard Smith and Szathmáry placed modern human behaviour, including creativity, into the broad context of evolutionary biology. Had Darwin been aware of genes and their role in evolution, he would perhaps have seen the parallels between genetic information and non-genetic information and reached similar conclusions. It should be emphasized that the emergence of *human language* is the first – and thus far the only – major transition identified by Maynard Smith and Szathmáry (1999: 149–170) that entailed changes in the way that the informational units of the brain (or *neuronal* information) are stored, transmitted, and translated.[2] And it should be noted that, despite the fact that Maynard Smith and Szathmáry stated the transition in terms of "primate societies to human societies," the emergence of language did *not* entail a transition to a eusocial entity or "super-organism" (see Hölldobler and Wilson, 2009), which is discussed below.

Neuronal information emerged among the metazoa more than half a billion years ago, and it is puzzling that so few organisms have evolved changes in how it is transmitted and translated (perhaps "innovations" in neuronal information did evolve among more organisms in the past, but lack visibility in the fossil record).[3] Both a specialized cell (neuron) and organ evolved more than 500 million years ago in the phylum Cnidaria (jellyfish, hydra, sea anemones, and corals) in order to collect and process information about the environment that could not be

---

2. It is probably not the last of such transitions, especially if true artificial intelligence or "strong AI" can be engineered in the near future (e.g., Kurzweil, 2005).
3. It is worth noting that non-human organisms that translate neuronal information to complex forms (see below) leave no potential fossil traces of their exotic brain functions, which undoubtedly would remain unknown and unsuspected had these taxa become extinct a million years ago.

inherited genetically (e.g., Allman, 1999; Swanson, 2003). The metazoan brain played a significant role in the explosion of complex life forms during the Cambrian, although it did not constitute a "major transition" in evolution.

In some respects, neuronal information – which the organism creates in the synaptic networks of its brain – is the reverse of genetic information. The latter is a design template for organisms, while the former is assembled from received sensory input. Both types of information are symbolic. While DNA codons are arbitrary referents or *symbols* for specific amino acids, various aspects of the environment – a chemical compound or the shape of a rock – are by necessity *represented* rather than reproduced in the brain among the networks of synaptic connections (Bickerton, 1990: 77–87). The structure and function of the latter allow storage of both digital (discrete or discontinuous) and analogical (continuous) units of information (e.g., LeDoux, 2002).

As the metazoa evolved, some taxa developed large phenotypes with bigger brains and more complex and diverse brain functions. In these taxa, increasingly complicated, hierarchically-organized, structures of neuronal information were generated in the brain. Representations based on visual sensory input became especially complex in certain groups, and it may be significant that humans evolved from animals with a highly sophisticated visual processing system (Marr, 1982; Hoffecker, 2011a: 39–41). A series of "innovations" in the visual system evolved among the early vertebrates (e.g., the optic tectum of the midbrain) and ancestral primates (e.g., expanded visual cortex, increased photoreceptor density in retina, enhanced colour vision) (e.g., Striedter, 2005: 301–310). The primates also evolved stereoscopic vision, creating three-dimensional representations in the brain.

Modern humans transform thought or neuronal information into three different forms. They translate neuronal information into *patterned arrangements* that exhibit a complex, hierarchically-organized structure, but lack information content, such as a musical composition or an abstract design.[4] They also translate information from the synaptic networks of the brain into *other forms of information*, such as a spoken sentence or a painting, in order to transmit the information to other brains. Finally, modern humans translate neuronal information into *artefacts* based on a complex hierarchically-structured design in a manner analogous to the transformation of genetic information into organisms.

Some of the same cognitive faculties also evolved in non-human species and these cases provide insights as to how and why they evolved in humans. A number of taxa have evolved a capacity for translating neuronal information to patterned arrangements that lack information, usually, if not exclusively, as courtship displays. They include various birds (e.g., nightingales), gibbons (*Hylobates*), and humpback whales (*Megaptera novaeangliae*) (Payne and McVay, 1971; Hauser, 1996; Clarke et al., 2006). Much less common is the faculty of translating neuronal information to other forms of information. Outside humans, it is known only in the honeybee (*Apis mellifera*), which transmits information concerning resource and potential nest site locations through a series of hierarchically-structured body movements (i.e., "dancing") (von Frisch, 1993; Seeley, 1995, 2010).[5] In this case, the faculty evolved as part of a strategy for foraging in less productive habitats (Seeley, 1985; Beekman and Lew, 2007). A capacity for translating neuronal information to complex, hierarchically-structured artifacts appears to be unique to humans, however.[6] Finally, it should be noted that the faculty of creative recombination of informational units evolved in humpback whales, apparently driven by sexual selection (Noad et al., 2000; Rothenberg, 2008: 131–168).

7.2. PAIR-BONDING AND COURTSHIP DISPLAYS

". . . some early progenitor of man probably first used his voice in producing true musical cadences, that is in singing, as do some of the gibbon-apes at the present day; and we may conclude from a wide-spread analogy, that this power would have been especially exerted during the courtship of the sexes."
Charles Darwin, *The Descent of Man and Selection in Relation to Sex* (1875: 87)

Charles Darwin articulated a theory of language (and music) origins in *The Descent of Man and Selection in Relation to Sex* that remains as plausible as any other. Although it received little attention for more than a century, his idea has been re-visited by several authors in recent years (e.g., Mithen, 2006: 178–182; Fitch, 2010: 397–399). Quite simply, Darwin (1875: 86–92) suggested that language (i.e., the translation of neuronal information to another form of information) emerged from complex courtship displays in the form of song (i.e., patterned arrangements devoid of

---

4. The phenomenon is nicely illustrated by Lewis Carroll's poem "Jabberwocky," a rare example of a patterned arrangement composed of written phonemes in the English language – exhibiting a complex, hierarchically-organized structure – that is devoid of information, despite the fact that it contains a number of recognizable words.

5. It should be noted that the brain of an individual honeybee contains only about one million neurons, although information-sharing and collective decision-making among thousands of honeybees creates a more powerful "cognitive entity" (Seeley, 2010: 198–217).
6. An isolated exception may be the New Caledonian crow, which is said to fashion simple digging tools from leaves in accordance with a "mental template" (Corballis, 2003: 168).

information). The latter eventually acquired referents and became information. The evolutionary context was sexual selection, which Darwin viewed as a significant factor in human evolution (Dawkins, 2004: 265–273).

It is logical to assume that the more complex forms of neuronal information translation in humans evolved from the simplest and most common form (i.e., patterned arrangements, such as courtship displays). Equally important, however, is the suggested link between the initial translation of neuronal information and sexual selection for courtship displays. This potentially ties it to a critical event in human evolution – the appearance of the long-term male–female pair bond (Chapais, 2008). Pair-bonding is generally uncommon among mammals and rare in the primates, although it is present in gibbons (Fuentes, 2000). It seems to promote a particularly intense form of sexual selection, and to yield highly elaborate courtship displays. And while displays based on generation of complex patterns from neuronal information might be expected in animals that evolved large brains and substantial computational faculties, they also may reflect the "runaway" effect of sexual selection (Darwin, 1875: 207–242; Dawkins, 2004: 265–270).

Living humans generate a variety of patterned arrangements that lack information, such as music, dance, and abstract designs (the patterns perceived by the observer are reconverted to neuronal information). These presumably have their roots in the courtship displays of early humans (and/or related displays of social solidarity outside the pair bond [see below]), which seem likely to have been confined to vocalizations and/or gestures. Living humans also frequently invest patterned arrangements with information, such as lyrics, which apparently reflects later developments in their evolution.

Although pair-bonding evolved in *Hylobates* and may therefore have been present in a human ancestor roughly 20 million years ago, it probably either re-evolved or evolved independently in humans following the divergence with the African Greater Apes. This conclusion is based on the fact that pair-bonding is not found in the latter, and also on the observation that earlier humans (australopithecines) exhibit a high degree of sexual dimorphism. Pair-bonding seems to select for sexual monomorphism in body size, and, outside modern humans, gibbons are the most monomorphic among higher primates (Fuentes, 2000). A significant reduction (∼15%) in sexual dimorphism among humans is not evident until roughly 1.8 million years ago (e.g., Klein, 2009; Mithen, 2006: 182–186).

Both reproductive factors and foraging demands probably drove the evolution of pair-bonding in humans. Long-term pair bonds were a solution to the increasingly high and protracted dependence of human infants and children, and the significant increase in cranial volume that immediately precedes the reduction of sexual dimorphism (e.g., Holloway et al., 2009) probably indicates added selection pressure for greater parental investment. Reduced sexual dimorphism also broadly coincides with the expansion into the temperate zone (∼2 million years ago) and colonization of less productive habitat (e.g., Gamble, 1994: 117–123). Pair-bonding and a sexual division of labour (both with respect to child care and food gathering) were part of the solution to foraging in environments where resources were widely dispersed.

Darwin (1875: 87) specifically mentioned the songs of gibbons in his discussion of the origin of music and language, and it should be noted that new research shows that not only are gibbon songs composed of complex, hierarchically-structured vocal sounds, but they also have acquired referents. In addition to courtship displays, gibbon songs function as alarm calls for terrestrial predators (leopard, tiger, and python) (Clarke et al., 2006: 4), similar to the predator alarm calls of vervet monkeys (Cheney and Seyfarth, 1990). This provides further support for Darwin's suggestion that language (i.e., transmission of information) developed from songs generated by translating neuronal information to complex patterned arrangements in the form of courtship displays.

7.3. FORAGING STRATEGY AND INFORMATION-SHARING

"... the basic plan of honeybee foraging involves the colony as an 'information center,' monitoring a vast area around the nest for food sources, pooling the reconnaissance of the foragers, and somehow using this information to focus a colony's forager force on a few, high-quality patches within its foraging range"
Thomas D. Seeley, *Honey Bee Ecology* (1985: 92)

If gibbons provide a model for the initial translation of neuronal information to another form outside the brain, the honeybee colony offers insights to why humans began to translate neuronal information to another form of information. Even among large-brained vertebrates, most animal communication is limited to simple bits of information concerning the affective state of the transmitter (Hauser, 1996: 473–522). As already noted, the honeybee is a rare exception, being able to communicate complex hierarchically-structured neuronal information pertaining to resource locations and nest-site characteristics to other honeybees in the same colony (von Frisch, 1993; Seeley, 1995; Riley et al., 2005). The collection and sharing of non-genetic information is an essential component of the honeybee foraging strategy – scouts cover a wide area and report back to the colony on resource locations – termed "information-center" foraging by Seeley (1985: 92). A similar strategy is implemented when a swarm gathers and evaluates data on a new nest site (Seeley, 2010).

The waggle dance has been compared to human language because it communicates information with arbitrary referents (and local "dialects" have emerged among various groups) to objects in another spatial-temporal setting (Bickerton, 1990: 153; Gould and Gould, 1995: 59–60; Hauser, 1996: 498–500). It is unique among non-human animals because it entails translation of neuronal information into another form (body movements) that allows transfer of a complex mental representation from one brain to others. The waggle dance differs from modern human language as a closed system, with genetically defined domains and no potentially unlimited creative recombination of informational units (Hauser, 1996: 496–504).

The honeybee colony offers a model for early human foraging and explanation of why humans began to collect and share information in a way that has limited precedence among the higher primates. The most complex information-sharing among non-human primates in the wild seems to be the predator-specific alarm calls of gibbons and vervet monkeys (Cheney and Seyfarth, 1990; Hauser, 1996: 509–513; Clarke et al., 2006). These do not, however, entail transmission of complex hierarchically-organized representations necessary for "information-centre" foraging.[7] Such a strategy is characteristic of recent modern human foraging peoples (e.g., Mithen, 1990; Kelly, 1995), and it seems likely to have been present in some form among early humans.

The most likely context for the emergence of an information-centre foraging strategy among early humans is the massive expansion of geographic range roughly 2 million years ago, when representatives of early *Homo* dispersed out of tropical Africa into the temperate zone of Eurasia (as far as latitude 40°N) (Gamble, 1994: 117–123; Klein, 2009). At this time, humans invaded a variety of northern habitats that were cooler, more seasonal, and less productive than those occupied by the australopithecines, while there is evidence for occupation of drier C_4 grasslands in sub-Saharan Africa (Plummer et al., 2009). Presumably, they adapted to environments in which resources were more widely dispersed in time, as well as in space.

Like the honeybee (which also expanded out of tropical Africa into the temperate zone [Gould and Gould, 1995: 24; Whitfield et al., 2006]), early humans may have adapted to these environments by evolving an information-centre strategy. Individual or small groups of foragers could have doubled as "scouts," spreading out across large areas and returning to a base camp with food and materials, as well as information on resource locations. The model assumes central-place foraging and large average patch size (to accommodate multiple foragers) (Hoffecker, 2012). There is consensus that the early human diet at this time was diverse – and many resources would have been concentrated in large patches – while at least some of the sites appear to reflect central-place foraging (Schick and Toth, 2006: 31).

There is, however, no direct evidence for the collection and sharing of information regarding resource (or potential camp) locations, and it is possible that other innovations, such as hunting or stone tool-making, permitted the expansion of geographic range.[8] The plausibility of the model lies primarily in the common assumption that syntactic language must have been preceded by a simpler system of information transmission (e.g., non-syntactic or "proto-language" [Bickerton, 1990]), and that the dispersal into temperate Eurasia 2 million years ago – preceded by the first significant increase in human brain volume (e.g., Holloway et al., 2009) – seems the most likely event to be linked to this development. The broad co-occurrence of sexual monomorphism and the colonization of less productive habitat suggests that pair-bonding and information-center foraging might have emerged at roughly the same time, and that, accordingly, faculties for translating neuronal information to patterned arrangements and other forms of information evolved at roughly the same time.

In one important respect, information-centre foraging among early humans would have differed from that of the honeybee colony. Honeybees are eusocial insects and a colony actually represents a very large nuclear family in which the high coefficient of relatedness among the foragers (i.e., siblings) promotes cooperation and information-sharing (i.e., based on kin selection [Hamilton, 1964; Maynard Smith, 1964]).[9] With respect to reproductive biology, humans lie at the opposite end of the spectrum, where the size of a nuclear family – especially among mobile foragers – is severely constrained by limited number of offspring and prolonged gestation and infant dependency. Because of marriage rules, information-sharing in recent foraging societies entailed cooperation among individuals with a comparatively low coefficient of relatedness (e.g., Hill et al., 2011).

If courtship songs composed of complex patterned arrangements of vocal sounds provided the basis for spoken language or proto-language, pair-bonding provided a basis

---

7. Nor do they refer to phenomena outside the immediate temporal-spatial setting of the individuals that transmit and receive the alarm calls, which are issued only in the presence of one of these predators (Hauser, 1996: 511; Clarke et al., 2006: 4).

8. As in the case of the courtship displays, neuronal information regarding resource locations or potential camp sites presumably was translated to another mode with vocalizations and/or gestures (Bickerton, 1990; Corballis, 2002).

9. In a recent paper co-authored by Edward O. Wilson, Nowak et al. (2010) argued that kin selection was not necessary to account for eusociality in the honeybee. This paper has generated considerable criticism and controversy among evolutionary biologists (e.g., Abbot et al., 2011), most of whom seem to view the kin selection model as the best explanation for eusociality.

for alliance networks composed largely of relationships outside the parent–offspring or sibling bond. Among recent hunter–gatherers, society is built upon reciprocal exogamous mate exchange, which creates an integrated network of exogamous pair-bonds (Chapais, 2008; Dunbar, 2010). In a sense, human society is an extension of the pair-bond, and it should generate a similar form of selection ("social selection"). A network based on long-term reciprocal relationships among non-relatives (or individuals outside the parent–offspring or sibling ties) should select for abilities to maximize the benefits over the costs of such relationships. As in the case of pair-bonding, there are opportunities for subtle competition and secret cheating (Trivers, 1971; Byrne and Whiten, 1988).

A major consequence of long-term reciprocal alliances (or "reciprocal altruism") is that it is not stable from an evolutionary perspective, and competition among the non-relatives within the system of alliances may drive evolutionary change (Trivers, 1971; Maynard Smith, 1982). Both sexual selection in the context of pair-bonding and "social selection" for networking skills may have been the primary force behind evolutionary change in the genus *Homo* after 2 million years ago. Between 2 and 0.5 million years ago, there was a steady and significant expansion in brain size – apparently without accompanying changes in foraging or technology – attributed to sexual and social competition ("social brain hypothesis") (Dunbar, 1996, 1998; Miller, 2000; Dawkins, 2004: 270). A byproduct of this expansion was a massive increase in the amount of storage capacity for neuronal information. But by 0.5 million years ago, the size of the human brain seems to have been approaching a maximum limit (especially for the tropical zone), and subsequent developments in *Homo* may reflect the effects of these selection pressures on other aspects of anatomy and behaviour.

7.4. THE EMERGENCE OF PHENOTYPIC THOUGHT

"The origin of language ... seems to have been closely linked with technical motor function. Indeed the link is so close that employing as they do the same pathways in the brain, the two ... could be attributed to one and same phenomenon."
André Leroi-Gourhan, *La Geste et la Parole* (1964: 165)

The appearance of artefacts composed of hierarchically-structured informational units is a major event in human evolution, not only because it seems unprecedented in biology, but because it established a pattern that most likely contributed to the subsequent emergence of creativity. By generating "phenotypes" of thought in the form of chipped stone artefacts, the human brain began to interact with its own informational units as pieces of the external environment to be seen, felt, and manipulated.[10] No wonder humans became conscious of their own thoughts (e.g., Hoffecker, 2011a: 58–60). Moreover, like the phenotype of an organism, the structures based on neuronal information were exposed to the effects of "selection" by environmental factors.

A faculty for translating neuronal information to complex artefacts may have evolved in humans for the same reason that the brain is widely thought to have expanded so dramatically between 2 and 0.5 million years ago. In this case, competition for mates and status within social networks apparently was played out in the public arena of tool making. It may have developed in a parallel fashion to that postulated for language or proto-language: the making of complex implements with functional properties (e.g., a hafted spear) probably had roots in the making of artefacts based on a patterned arrangement without functional properties (hand axes?).

In retrospect, tool-making seems to be an inevitable medium for the expression of sexual and social competition among humans because it had become a uniquely important part of their lives. Chimpanzees make and use simple tools and it is conceivable that the last common ancestor of chimpanzees and humans was a tool-maker. In any case, the shift to bipedal locomotion that apparently marked the divergence of the two lineages seems to have initiated a trend towards increased specialization of the forelimbs. Between 3.5 and 1.5 million years ago, there are several changes in the anatomy and function of the hand (e.g., expanded length of thumb [Marzke, 1983]), and by the time that the earliest hand axes appear, humans had evolved a highly sensitive and precise instrument for manipulating objects (Napier, 1993; Mountcastle, 2005). The hand acquired a function analogous to that of the vocal tract, and the multiplicity of subtle movements and grips eventually manifested the same potential for an infinite range of hierarchically-structured combinatorial variations (Hoffecker, 2011a: 48–51).

It seems more than likely that early hominins were making and using a variety of tools similar, if not superior, to those of modern chimpanzees, but the earliest known stone tool assemblages (Oldowan) date to no more than 2.6 million years ago (Schick and Toth, 2006). They reflect a capacity for flaking control that chimpanzees lack (Toth et al., 1993), but do not exhibit the imposition of a mental template or design on the rock. The Oldowan tools appear to be entirely reductive (including the "spheroids"), and the shape of the finished implement is similar to that of the

---
10. The subject has been addressed by Martin Heidegger and other philosophers of technology (see Mitcham and Mackey, 1983), and in recent years, by archaeologists under the heading "material engagement of the mind" (Renfrew, 2004). It also is germane to the concept of the "distributed mind" (e.g., Dunbar et al., 2010).

original cobble or flake from which it was fashioned (Toth, 1985; Schick and Toth, 1993; Mithen, 1996: 96–98; Wynn, 1999: 264–268).

Artefacts classified as hand axes or bifacial tools currently are dated as early as 1.76 million years ago (Lepre et al., 2011) and, for at least a few hundred thousand years, are confined to sub-Saharan Africa. While archaeologists traditionally have recognized the hand axe as a "type" comprising several "sub-types," its emergence and development in the archaeological record has a gradational or blurred character that presumably reflects a mentality somewhat different from that of modern humans. To begin with, formal bifaces are preceded by a lengthy history of partial or complete bifacial flaking of various artifacts in the Oldowan (see Leakey, 1971). The earliest recognized bifaces are rather crudely – and sometimes only partially – flaked, and similar to the "proto-bifaces" of the Oldowan (Wynn, 1991: 199–203). Later bifaces contain more finely shaped and symmetrical specimens, but many exhibit little or no symmetry and a minimum of thinning and shaping (e.g., McNabb et al., 2004: 664–666). Moreover, the various sub-types represent a continuum of variation (Isaac, 1977: 116–145; McPherron, 2006).

Despite the lack of a well defined design comparable to the material culture of modern humans (e.g., Folsom fluted point), hand axes are widely perceived to reflect the imposition of a "mental template" on a piece of stone (e.g., Schick and Toth, 1993: 237–245; Wynn, 1995; Mithen, 1996: 117–119; Pelegrin, 2009: 100–102).[11] In contrast to the Oldowan artifacts (Toth, 1985), the finished biface bears a limited relationship to the original piece of rock (although it may bear a recognizable relationship to the blank struck from that rock [e.g., McNabb et al., 2004: 664]). A preconceived design would seem to have been necessary to produce many of the later bifaces, which entailed three irreversible, hierarchically-organized steps: (1) striking a large flake blank off a core; (2) flaking both sides of the blank to create an ovate form in three dimensions; and (3) trimming the edges to render a more evenly shaped artefact (Gowlett, 1984: 180, Fig. 7.3; Schick and Toth, 1993: 237–245; Wynn, 1995; Pelegrin, 2009: 100–102).

Why did African *Homo* begin making hand axes 1.8–1.7 million years ago? The appearance of bifacial stone artefacts generally has been explained in functional terms, and there is some supporting microwear evidence for their use as butchering tools (Keeley, 1993; Schick and Toth, 1993: 258–260; Pitts and Roberts, 1998: 285–287). However, many bifaces exhibit little or no trace of use (e.g., Jelinek, 1977) and they played no role in the geographic expansion of early humans into the temperate zone – Oldowan tools were adequate for their needs. Hand axes are rare in East Asia, where early *Homo* apparently occupied a similar ecological niche without them (Boaz and Ciochon, 2004: 95–107). In my view, the functional arguments are weak,[12] and their role as tools may have been a secondary one.

At present, the most parsimonious explanation of the hand axes is that they represent a patterned arrangement based on a design that lacks both information content and functional properties. They are plausibly interpreted as an artefactual form of courtship display (Kohn and Mithen, 1999; Mithen, 2006: 188–191) and/or a form of display within the wider context of the alliance networks that probably had evolved by this time. In any case, they are the logical source of the subsequent translation of neuronal information to complex artefacts based on a hierarchically-structured production sequence and a design that exhibits functional properties. These included composite tools and weapons fitted with stone blanks that were produced with a novel prepared-core technique to yield a specific size and shape.

New research reveals that composite implements were being made by 0.5 million years ago (Wilkins et al., 2012), which coincides with the appearance of prepared-core techniques (Klein, 2009). They comprised three or four components: (1) handle/shaft, (2) blade/point, (3) binding cord, and (4) adhesive. Each component was manufactured from a different raw material (i.e., wood, stone, hide, and some material or compound with adhesive properties such as resin) and processed in a different way (Anderson-Gerfaud, 1990; Boëda et al., 1996; Lombard, 2005, 2007; Barham, 2010). The components were brought together in a hierarchically-structured, preconceived design that seems to have had some combinatorial variations (e.g., side-blade versus end-blade) (Ambrose, 2001: 1751; Villa et al., 2010). In contrast to the hand axes, there can be little doubt that these artefacts contributed to the fitness of their makers (*Homo heidelbergensis*) by improving the power and efficiency of their tools and weapons.[13]

---

11. As Wynn (1999: 268) notes, "the knappers imposed an arbitrary shape on the tool," but several archaeologists have objected to application of the term "mental template" as either too prescribed or altogether inappropriate, given the lack of standardization (e.g., Noble and Davidson, 1996: 200). Gowlett (2006: 205) wrote that a *mental template* "would yield neither the fields of variation within a dataset, nor the local variations observed from site to site and even within sites." McNabb et al. (2004: 667) suggested that a more appropriate label would be "individualized memic constructs" although some may lack a mental template for this term.

12. Wynn (1999: 273) concluded that "functional explanations of bifaces have been largely unsatisfactory."
13. Evidence for composite implements is based on the stone components (with traces of hafting microwear and adhesives); equally complex technology in wood, hide, and soft plant materials seems likely, but has little or no archaeological visibility in this time range with rare exceptions (e.g., Haidle, 2009).

The appearance of complex artefacts marked a turning point in human evolution. For the first time, humans (or at least one group of humans) had evolved a cognitive faculty that was and remains unique to the genus *Homo*, and one that probably was critical to the emergence of the modern human mind. The translation of neuronal information to complex technology seems to have set the stage for the subsequent evolution of a faculty for creatively recombining informational units into a potentially infinite variety of arrangements and structures (Ambrose, 2010; Hoffecker, 2011a).

7.5. THE ARCHAEOLOGY OF ALTERNATIVE REALITY

"You don't necessarily need the prefrontal cortex to form the mental image of a human or of a fish, but you need it to form the mental representation of a mermaid"
Elkhonon Goldberg, *The New Executive Brain* (2009: 34)

Anatomically speaking, modern humans are present in sub-Saharan Africa by 200,000 years ago (Willoughby, 2007), but evidence for fully modern behaviour currently dates no earlier than 75,000 years ago (Henshilwood, 2007; Tattersall, 2009). The apparent discrepancy may simply reflect the limited archaeological visibility of evidence for such behavior – problems of sampling and preservational bias – and earlier evidence may turn up in the next few years.

Modern behaviour or *modernity* is often equated with the use of symbols or more explicitly in archaeological terms, with signs of the "material storage of symbols" in the form of abstract notation or art objects (Donald, 1991; Renfrew, 1998; Wadley, 2001; Henshilwood and Marean, 2003). This, in turn, is widely equated with syntactic language, although there is no direct archaeological evidence for the latter until a few thousand years ago. I would argue, however, that information in the form of symbols is only part of what modern humans do with neuronal information; they also transform it into patterned arrangements that may have no referents (i.e., not symbolic) and technology based on hierarchically-structured, often highly complex, design. The essential difference between the modern human mind and its predecessors is that the former exhibits a capacity for potentially unlimited recombination of informational units or *creativity* (Corballis, 2011; Hoffecker, 2011a: 73–77).

The anatomical change in later *Homo* evolution that may be the most important for the appearance of creativity is the expansion of the prefrontal cortex in the frontal lobe (Kornhuber, 1993), on which "the ability to manipulate and recombine internal representations critically depends" (Goldberg, 2009: 23). Modern humans exhibit at least a modest expansion of this part of the brain (Holloway, 2002), which is implicated in the performance of novel tasks, according to brain-imaging studies (Goldberg, 2009: 89–91), and also in working memory capacity (Coolidge and Wynn, 2005). More generally, the substantial increases in cranial volume – providing for exponential growth of synaptic networks – that took place during the several hundred thousand years preceding the appearance of anatomically modern humans laid the foundation for creativity by accommodating an immense store of information (Holloway, 1995; Holloway et al., 2009).

The archaeological evidence for creativity comprises a body of artefacts and features that illustrate combinatorial variation of elements within a hierarchical structure with multiple levels and nested components (Hoffecker, 2007: 377–378, 2011a: 100–103). The categories of artefacts and features include all three forms of neuronal information translation: (1) to patterned arrangements that probably are devoid of information such as abstract designs on objects or cave walls and, by implication, music compositions (based on the recovery of wind instruments); (2) to artefacts that exhibit a complex, hierarchically-structured design such as, by implication, tailored clothing (based on the recovery of eyed needles); and (3) to other forms of information, such as visual art in two or three dimensions. Collectively, these artifacts provide analogues to narratives generated from a syntactic language and suggest that the potential number of variations is infinite. Isolated artifacts – regardless of the complexity of their organization – are insufficient because they do not demonstrate combinatorial variations.

The best illustrations of creativity are found in the visual art of the early Upper Paleolithic in Europe. The oldest specimen is what appears to be the head fragment of a human figurine carved in mammoth ivory, recovered from below the 40,000-year-old volcanic tephra at Kostenki 14 on the Don River in Russia, and probably dating to about 44,000–42,000 cal BP (Sinitsyn, 2002: 230, Fig. 9; Anikovich et al., 2007). The facial features are missing (which is why it remains somewhat problematic as a sculpture) and the piece may have been abandoned before completion.[14]

The critical body of evidence is slightly younger (~40,000–35,000 cal BP) and recovered from rock shelters in Western Europe. It includes the two-dimensional images from Chauvet Cave in southern France, and the human and animal figurines from southern Germany and Austria (e.g., Hahn, 1972; Marshack, 1990; Mithen, 1996; Clottes, 2003; Conard, 2003). These are artificial visual representations –

---

14. The possible ivory figurine head was recovered from a rich assemblage containing backed bladelets, non-stone artifacts (e.g., antler mattocks), and materials imported from distances of >500 km (Anikovich et al., 2007), that I have suggested may be at least tentatively included in an East European variant of the Proto-Aurignacian industry (Hoffecker, 2011b).

created in the brain and projected into the external world and to other brains – analogous to other forms of information such as spoken language. They contain multiple nested components and these may be seen to vary from one representation to another (e.g., compare the figurines from Hohle Fels Cave and Hohlenstein-Stadel [Marshack, 1990: 479, Fig. 17.17; Conard, 2009: 248, Fig. 1]); the potential variations would seem to be unlimited. The Hohlenstein-Stadel *Löwenmensch* figurine is particularly striking because it depicts an organism that combines the features of a human with that of another mammal.

The evidence for creativity in technology emerges with less drama in the archaeological record. This is partly because the more complex forms of technology are inferred – often represented at best by an isolated fragment or two. Fragments of eyed needles from an early Upper Paleolithic cave in the northern Caucasus (Golovanova et al., 2010) indicate sewn clothing at 40,000 cal BP, which – if similar to the sewn clothing of non-industrial peoples at high latitudes – would have been a very complicated piece of technology with numerous components (Hoffecker, 2005: 188; Gilligan, 2010). Devices for snaring and/or trapping small mammals may be inferred from the concentration of hare remains in a pre-40,000-year-old layer at Kostenki 14, but the design of the technology is unknown (Hoffecker, 2005: 189). On the other hand, examples of probable artificial memory systems (or *information technology*) are represented in early Upper Paleolithic sites in Western Europe that antedate 30,000 cal BP (e.g., Abri Blanchard, France (Marshack, 1972; d'Errico, 1998)).

Before 45,000 years ago, archaeological evidence for creativity in both visual art (information) and technology is limited (while evidence for music is entirely lacking). The pattern either indicates that the capacity for creative recombination of neuronal information emerged gradually over an extended period of time (e.g., McBrearty and Brooks, 2000), or simply reflects the reduced archaeological visibility of the earlier record – especially given the likelihood that much of the evidence for creativity is less easily preserved (e.g., bone artefacts). In either case, it appears almost certain that the capacity for potentially unlimited generation of variation in both informational and technological structure had evolved by the time that anatomically modern humans dispersed out of Africa (roughly 60,000–50,000 years ago). This is because it is unlikely that syntactic language and other forms of unlimited creative expression evolved more than once, and also because the ability to devise novel and often complex technologies was a prerequisite for the rapid global dispersal (e.g., Hoffecker, 2005: 187–190).

Ostrich eggshell containers were decorated with simple geometric designs at Diepkloof Rock Shelter (South Africa) at about 60,000 years ago (Texier et al., 2010). A 75,000-year-old occupation layer at Blombos Cave (also South Africa) contains red ochre pieces engraved with simple geometric designs, as well as bone tools and perforated shell ornaments (Henshilwood et al., 2001; Henshilwood, 2007). The geometric designs inscribed on shell and ochre may have had some meaning (i.e., contained a message), but they seem equally likely to represent a visual analogue to music – neuronal information translated to a patterned arrangement that is devoid of information.

Perforated shell ornaments have been recovered from even earlier contexts in North Africa and the Levant, dating to as much as 135,000 years ago (Vanhaeran et al., 2006). The ornaments may have been worn as individual pieces or might have been combined into necklaces comprising many components (i.e., entailing a number of hierarchically-organized steps to assemble). They may have been worn to signal social and/or ethnic identity, and thus played a role analogous to that of the hand axes. Evidence for creativity in technology before 45,000 years ago also includes several barbed bone points, dating at 90,000 years ago, recovered at Katanda in Zaire (Yellen et al., 1995).

Why did a faculty for creative recombination of informational units evolve in later *Homo*, and why did it evolve among the African *Homo* population,[15] which subsequently dispersed to other parts of the earth (Gamble, 1994; Klein, 2009)? A similar faculty for creativity also evolved in the humpback whale, apparently as a consequence of sexual selection (Noad et al., 2000; Rothenberg, 2008). This suggests that the same forces believed to lie behind the expansion of the brain and possibly the appearance of composite implements may be implicated in creativity, but it also indicates that creative recombination of neuronal information could evolve in the absence of the unique human faculty for translating it to complex artefacts. Whales only evolved a capacity for the recombination of elements in the category of patterned arrangements ("songs"), however, while humans developed a faculty that was applied to all categories of translation. Moreover, the capacity for translating novel arrangements of neuronal information to other forms of information yielded syntactic language and visual art, allowing a collective form of creative recombination and the emergence of a generative "super-brain" (Hoffecker, 2012).

Explaining the evolution of modern behaviour or creativity in sub-Saharan Africa – as opposed to other parts of the world inhabited by descendents of *Homo*

---

15. Some evidence for creativity in Western Europe is attributed to the local Neanderthal population (e.g., Riel-Salvatore, 2010), but in my view it is problematic, either because it is represented by artefacts that may have been produced by modern humans but mechanically mixed with Neanderthal skeletal remains (e.g., Bordes and Teyssandier, 2012) or by industries that are widely assigned to the Neanderthals, but may actually have been produced by modern humans (e.g., Adams, 2007).

heidelbergensis – remains a challenge. One explanation is that creativity (or at least syntactic language) was the result of a random mutation ("neural hypothesis") that might have occurred in any of the large-brained *Homo* species in Africa and Eurasia, and took place in the former only by chance (Klein, 2009: 647–649). The "information environment" might have been a factor, however, in lower latitudes. *Homo* populations in sub-Saharan Africa, as well as those living in southern Asia, inhabited places characterized by a significantly higher biodiversity than their counterparts at higher latitudes (especially during cold periods).[16] They may also have lived at higher population densities (more persons per unit area) owing to greater carrying capacity. Given the substantial increases in cranial volume after 1.8 million years ago with their implications for increased storage of neuronal information, the African population may have been exposed to *information overload*, especially as brain size approached a maximum limit. Experimental studies indicate that very large quantities of information – even when relevant to a specific problem – inhibit decision-making among modern humans (see Gleick, 2011: 405–412). An important function of syntactic language – with its flexible hierarchical structure – is efficient organization of large amounts of information.

7.6. CONCLUSIONS

Humans evolved the capacity for creativity (or generativity) in the context of multiple faculties for the translation of neuronal information to other forms outside the brain. Earlier humans probably evolved a faculty for translating neuronal information to patterned arrangements devoid of information in the form of vocalizations and/or gestures (i.e., courtship displays) in the context of pair-bonding before 1.8 million years ago. At roughly the same time, they may also have evolved a faculty for translating neuronal information to one or more other forms of information (e.g., "proto-language") in the context of an "information-centre" foraging strategy that permitted colonization of less productive habitats by reciprocal alliance networks composed of exogamous pair-bonds. Darwin (1875) suggested that translation to other forms of information (i.e., language) probably was derived from the patterned arrangements without information (specifically song).

---

16. Species diversity in most, if not all, of the habitats occupied by modern humans in the African tropical zone 250,000–75,000 years ago would have exceeded that of habitats occupied by the European Neanderthals by several orders of magnitude. For example, a four-acre plot in a tropical rain forest may contain up to 227 species of trees, while the same area in a deciduous forest in Michigan will contain only 10–15 species of trees; while there are an estimated 500–600 species of land birds in Central America, there are less than 150 species in the North American Midwest (Krebs, 1978: 458–459).

Despite a long history of tool-making, there is no evidence that humans had evolved a faculty for the translation of neuronal information into technological form until relatively late (~0.5 million years ago), when they began to produce composite tools and weapons that reflect a complex, hierarchical organization. This faculty also probably evolved from the patterned arrangements and is plausibly attributed to the same social and sexual selection pressures created by the reciprocal alliance networks and their component pair-bonds that are widely believed to underlie the increase in brain size 1.5–0.5 million years ago (Dunbar, 1996). Its timing may be related to constraints on further brain expansion after 0.5 million years ago. Neuronal design-based technology is unique to humans, and it established a relationship between a metazoan brain and mutable arrangements of translated information units outside the brain ("external thoughts"). This dynamic relationship, which existed within the unstable evolutionary setting of the alliance networks and pair-bonds, probably was the source for the cognitive faculty of generativity (potentially unlimited recombination of information units within a hierarchical structure) for which there is archaeological evidence after 0.1 million years ago.

Both the translation of neuronal information to complex artefacts and the faculty for potentially infinite recombination of information units represent "major transitions" in evolution with respect to non-genetic information, as defined by Maynard Smith and Szathmáry (1995, 1999), analogous to the origin of prokaryotes and sexual reproduction, respectively. In each case, the new forms of information could replicate only as a part of a larger whole and, following these transitions, translated neuronal information begins to exhibit its own pattern of evolutionary change. A third major transition – artificial intelligence or AI – appears imminent (Kurzweil, 2005) and probably will be attributable to the same selection pressures that drove the previous transitions. Unlike the latter, AI will emerge without associated changes in genetic information.

ACKNOWLEDGEMENTS

I am grateful to Scott A. Elias, who invited me to contribute this chapter following a talk on "human evolution and information" presented at Royal Holloway, University of London in November 2011. I also thank G. Richard Scott, who provided comments on a similar talk presented in October 2011 at the University of Nevada-Reno; Steven L. Kuhn, who commented on the draft; and Valerie E. Stone for stimulating and fruitful discussions on many of the issues addressed in this chapter.

REFERENCES

Abbot, P., et al., 2011. Inclusive fitness theory and eusociality. Nature 471, E1–E4.

Adams, B., 2007. Gulyás archaeology: the Szeletian and the Middle to Upper Palaeolithic transition in Hungary and central Europe. In:

Riel-Salvatore, J., Clark, G.A. (Eds.), New Approaches to the Study of Early Upper Paleolithic 'Transitional' Industries in Western Eurasia: Transitions Great and Small, pp. 91–110. BAR International Series 1620.

Allman, J.M., 1999. Evolving Brains. Scientific American Library, New York.

Ambrose, S.H., 2001. Paleolithic technology and human evolution. Science 291, 1748–1753.

Ambrose, S.H., 2010. Coevolution of composite-tool technology, constructive memory, and language: implications for the evolution of modern human behavior. Curr. Anthropol. 51 (Suppl. 1), S135–S147.

Anderson-Gerfaud, P., 1990. Aspects of behavior in the Middle Palaeolithic: functional analysis of stone tools from Southwest France. In: Mellars, P. (Ed.), The Emergence of Modern Humans. Edinburgh University Press, Edinburgh, pp. 389–418.

Anikovich, M.V., et al., 2007. Early Upper Paleolithic in Eastern Europe and implications for the dispersal of modern humans. Science 315, 223–226.

Barham, L., 2010. A technological fix for 'Dunbar's Dilemma'? Proc. Br. Acad. 158, 367–389.

Beekman, M., Lew, J.B., 2007. Foraging in honeybees—when does it pay to dance? Behav. Ecol. 255–262.

Bickerton, D., 1990. Language & Species. University of Chicago Press, Chicago.

Boaz, N.T., Ciochon, R.L., 2004. Dragon Bone Hill: An Ice Age Saga of *Homo erectus*. Oxford University Press, Oxford.

Boëda, E., et al., 1996. Bitumen as a hafting material on Middle Palaeolithic artefacts. Nature 380, 336–338.

Bordes, J-G., Teyssandier, N. 2012. The Upper Paleolithic nature of the Chatelperronian in South-Western France: archeostratigraphic and lithic evidence. Quaternary International 259, 95–101.

Byrne, R., Whiten, A. (Eds.), 1988. Machiavellian Intelligence. Oxford University Press, Oxford.

Chapais, B., 2008. Primeval Kinship: How Pair-bonding Gave Birth to Human Society. Harvard University Press, Cambridge.

Cheney, D.L., Seyfarth, R.S., 1990. How Monkeys See the World: Inside the Mind of Another Species. University of Chicago Press, Chicago.

Childe, V.G., 1956. Society and Knowledge: The Growth of Human Traditions. Harper & Brothers, New York.

Chomsky, N., 1988. Language and the Problems of Knowledge: The Managua Lectures. MIT Press, Cambridge, Mass.

Chomsky, N., 2002. On Nature and Language. Cambridge University Press, Cambridge.

Clarke, E., Reichard, U.H., Zuberbühler, K., 2006. The syntax and meaning of wild gibbon songs. PLoS One 1 (1), e73.

Clottes, J., 2003. Chauvet Cave: The Art of Earliest Times. University of Utah Press, Salt Lake City.

Conard, N.J., 2003. Palaeolithic ivory sculptures from southwestern Germany and the origins of figurative art. Nature 426, 830–832.

Conard, N.J., 2009. A female figurine from the basal Aurignacian of Hohle Fels Cave in Southwestern Germany. Nature 459, 248–252.

Conard, N.J., Malina, M., Münzel, S.C., 2009. New flutes document the earliest musical tradition in southwestern Germany. Nature 460, 727–740.

Coolidge, F., Wynn, T., 2005. Working memory, its executive functions, and the emergence of modern thinking. Camb. Archaeol. J. 15, 5–26.

Corballis, M.C., 2002. From Hand to Mouth: The Origins of Language. Princeton University Press, Princeton.

Corballis, M.C., 2003. Recursion as the key to the human mind. In: Sterelny, K., Fitness, J. (Eds.), From Mating to Mentality: Evaluating Evolutionary Psychology. Psychology Press, New York, pp. 155–171.

Corballis, M.C., 2011. The Recursive Mind: The Origins of Human Language, Thought, and Civilization. Princeton University Press, Princeton.

Darwin, C.R., 1875. The Descent of Man and Selection in Relation to Sex. John Murray, London.

Dawkins, R., 1976. The Selfish Gene. Oxford University Press, New York.

Dawkins, R., 2004. The Ancestor's Tale: A Pilgrimage to the Dawn of Evolution. Houghton Mifflin, Boston.

d'Errico, F., 1998. Palaeolithic origins of artificial memory systems: an evolutionary perspective. In: Renfrew, C., Scarre, C. (Eds.), Cognition and Material Culture: The Archaeology of Symbolic Storage. McDonald Institute Monographs, Cambridge, pp. 19–50.

d'Errico, F., et al., 2003. Archaeological evidence for the emergence of language, symbolism, and music – an alternative multidisciplinary perspective. J. World Prehist. 17, 1–70.

Donald, M., 1991. Origins of the Modern Mind: Three Stages in the Evolution of Culture and Cognition. Harvard University Press, Cambridge.

Dunbar, R.I.M., 1996. Grooming, Gossip, and the Evolution of Language. Harvard University Press, Cambridge.

Dunbar, R.I.M., 1998. The social brain hypothesis. Evol. Anthropol. 6, 178–190.

Dunbar, R., 2010. Deacon's dilemma: the problem of pair-bonding in human evolution. Proc. Br. Acad. 158, 155–175.

Dunbar, R., Gamble, C., Gowlett, J. (Eds.), 2010. Social Brain, Distributed Mind. Oxford University Press, Oxford. Proceedings of the British Academy, No. 158.

Edelman, G.M., 2004. Wider than the Sky: The Phenomenal Gift of Consciousness. Yale University Press, New Haven.

Fine, C. (Ed.), 2008. The Britannica Guide to the Brain. Constable and Robinson Ltd, London.

Fitch, W.T., 2010. The Evolution of Language. Cambridge Univ. Press, Cambridge.

Fitch, W.T., Hauser, M.D., 2004. Computational constraints on syntactic processing in a nonhuman primate. Science 303, 377–380.

Fuentes, A., 2000. Hylobatid communities: changing views on pair bonding and social organization in hominoids. Yearb. Phys. Anthropol. 43, 33–60.

von Frisch, K., 1993. The Dance Language and Orientation of Bees. Harvard University Press, Cambridge.

Gamble, C., 1994. Timewalkers: The Prehistory of Global Civilization. Harvard University Press, Cambridge.

Gilligan, I., 2010. The prehistoric development of clothing: archaeological implications of a thermal model. J. Archaeol. Meth. Theor. 17, 15–80.

Gleick, J., 2011. The Information: A History, A Theory, A Flood. Pantheon Books, New York.

Goldberg, E., 2009. The New Executive Brain: Frontal Lobes in a Complex World. Oxford University Press, Oxford.

Golovanova, L.V., Doronichev, V.B., Cleghorn, N.E., Kulkova, M.A., Sapelko, T.V., Shackley, M.S., 2010. Significance of ecological factors in the Middle to Upper Paleolithic transition. Curr. Anthropol. 51, 655–691.

Gould, J.L., Gould, C.G., 1995. The Honey Bee. Scientific American Library, New York.

Gowlett, J.A.J., 1984. Mental abilities of Early Man: a look at some hard evidence. In: Foley, R. (Ed.), Hominid Evolution and Community Ecology: Prehistoric Human Adaptation in Biological Perspective. Academic Press, London, pp. 167–192.

Gowlett, J.A.J., 2006. The elements of design form in Acheulian bifaces: modes, modalities, rules and language. In: Goren-Inbar, N., Sharon, G. (Eds.), Axe Age: Acheulian Tool-making from Quarry to Discard. Equinox Publishing Ltd, London, pp. 203–221.

Hahn, J., 1972. Aurignacian signs, pendants, and art objects in central and Eastern Europe. World Archaeol. 3, 252–266.

Haidle, M.N., 2009. How to think a spear. In: de Beaune, S.A., Coolidge, F.L., Wynn, T. (Eds.), Cognitive Archaeology and Human Evolution. Cambridge University Press, Cambridge, pp. 57–73.

Hamilton, W.D., 1964. The genetical evolution of social behavior. I. J. Theor. Biol. 7, 1–16.

Hauser, M.D., 1996. The Evolution of Communication. The MIT Press, Cambridge.

Hauser, M.D., Chomsky, N., Fitch, W.T., 2002. The faculty of language: what is it, who has it, and how did it evolve? Science 298, 1569–1579.

Henshilwood, C., 2007. Fully Symbolic sapiens behaviour: innovation in the Middle Stone Age at Blombos Cave, South Africa. In: Mellars, P., Boyle, K., Bar-Yosef, O., Stringer, C. (Eds.), Rethinking the Human Revolution: New Behavioural and Biological Perspectives on the Origin and Dispersal of Modern Humans. McDonald Institute for Archaeological Research, Cambridge, pp. 123–132.

Henshilwood, C.S., D'Errico, F., Marean, C.W., Milo, R.G., Yates, R., 2001. An early bone tool industry from the Middle Stone Age at Blombos Cave, South Africa: implications for the origins of modern human behaviour, symbolism and language. J. Hum. Evol. 41 (6), 631–678.

Henshilwood, C., Marean, C.W., 2003. The origin of modern human behavior: critique of the models and their test implications. Curr. Anthropol. 44, 627–651.

Hill, K.R., et al., 2011. Co-residence patterns in hunter-gatherer societies show unique human social structure. Science 331, 1286–1289.

Hoffecker, J.F., 2005. Innovation and technological knowledge in the Upper Paleolithic of northern Eurasia. Evol. Anthropol. 14, 186–198.

Hoffecker, J.F., 2007. Representation and recursion in the archaeological record. J. Archaeol. Meth. Theor. 14 (4), 370–375.

Hoffecker, J.F., 2011a. Landscape of the Mind: Human Evolution and the Archaeology of Thought. Columbia University Press, New York.

Hoffecker, J.F., 2011b. The early Upper Paleolithic of Eastern Europe reconsidered. Evol. Anthropol. 20 (1), 24–39.

Hoffecker, J.F., 2012. The information animal and the super-brain. J. Archaeol. Meth. Theor. published online 12/17/11.

Hölldobler, B., Wilson, E.O., 2009. The Super-Organism: The Beauty, Elegance, and Strangeness of Insect Societies. W. W. Norton, New York.

Holloway, R.L., 1969. Culture: A human domain. Curr. Anthropol. 10, 395–412.

Holloway, R.L., 1995. Evolution of the human brain. In: Lock, A., Peters, C.R. (Eds.), Handbook of Human Symbolic Evolution. Clarendon Press, Oxford, pp. 74–125.

Holloway, R.L., 2002. Brief communication: how much larger is the relative volume of Area 10 of the prefrontal cortex in humans? Am. J. Phys. Anthropol. 118, 399–401.

Holloway, R.L., Sherwood, C.S., Hof, P.R., Rilling, J.K., 2009. Evolution of the Brain in Humans—Paleoneurology. In: Binder, M.D., Hirokawa, N., Windhorst, U. (Eds.), Encyclopedia of Neuroscience. Springer, New York, pp. 1326–1334.

Isaac, G. Ll, 1977. Olorgesailie: Archeological Studies of a Middle Pleistocene Lake Basin in Kenya. University of Chicago Press, Chicago.

Jelinek, A.J., 1977. The Lower Paleolithic: current evidence and interpretation. Annu. Rev. Anthropol. 6, 11–32.

Keeley, L.H., 1993. Microwear analysis of lithics. In: Singer, R., Gladfelter, B.G., Wymer, J.J. (Eds.), The Lower Paleolithic Site at Hoxne, England. University of Chicago Press, Chicago, pp. 129–138.

Kelly, R.L., 1995. The Foraging Spectrum: Diversity in Hunter-Gatherer Lifeways. Smithsonian Institution Press, Washington.

Klein, R.G., 2009. The Human Career: Human Biological and Cultural Origins, third ed. University of Chicago Press, Chicago.

Kohn, M., Mithen, S.J., 1999. Handaxes: products of sexual selection? Antiquity 73, 518–526.

Kornhuber, H.H., 1993. Prefrontal cortex and *Homo sapiens* – on creativity and reasoned will. Neurol. Psychiatr. Brain Res. 2, 1–6.

Krebs, C.J., 1978. Ecology: The Experimental Analysis of Distribution and Abundance, second ed. Harper & Row, New York.

Kurzweil, R., 2005. The Singularity is Near: When Humans Transcend Biology. Viking, New York.

Kyriacou, A., 2009. Innovation and creativity: a neuropsychological perspective. In: de Beaune, S.A., Coolidge, F.L., Wynn, T. (Eds.), Cognitive Archaeology and Human Evolution. Cambridge University Press, Cambridge, pp. 15–24. 2009.

Leakey, M.D., 1971. Olduvai Gorge. In: Excavations in Beds I and II, 1960–1963, vol. 3. Cambridge University Press, Cambridge.

LeDoux, J., 2002. Synaptic Self: How Our Brains Become Who We Are. Penguin Books, New York.

Lepre, C.J., Roche, H., Kent, D.V., Harmand, S., Quinn, R.L., Brugal, J.-P., Texier, P.-J., Lenoble, A., Feibel, C.S., 2011. An earlier origin for the Acheulian. Nature 477, 82–85.

Leroi-Gourhan, A., 1964. Le Geste et la Parole I: Technique et Langage. Albin Michel, Paris.

Leroi-Gourhan, A., 1965. Le Geste et la Parole II: La Mémoire et les Rythmes. Albin Michel, Paris.

Lombard, M., 2005. Evidence of hunting and hafting during the Middle Stone Age at Sibudu Cave, KwaZulu-Natal: A Multianalytical Approach. J. Hum. Evol. 48, 279–300.

Lombard, M., 2007. The gripping nature of ochre: the association of ochre with Howiesons Poort adhesives and Later Stone Age mastics from South Africa. J. Hum. Evol. 53, 406–419.

Marr, D., 1982. Vision: A Computational Investigation into the Human Representation and Processing of Visual Information. W. H. Freeman and Co, San Francisco.

Marshack, A., 1972. The Roots of Civilization: The Cognitive Beginnings of Man's First Art, Symbol and Notation. McGraw-Hill, New York.

Marshack, A., 1990. Early hominid symbol and evolution of the human capacity. In: Mellars, P. (Ed.), The Emergence of Modern Humans. Edinburgh University Press, Edinburgh, pp. 457–498.

Marzke, W.M., 1983. Joint function and grips of the Australopithecus afarensis hand, with special reference to the region of the capitate. J. Hum. Evol. 12, 197–211.

Maynard Smith, J., 1964. Group selection and kin selection: a rejoinder. Nature 201, 1145–1147.

Maynard Smith, J., 1982. Evolution and the Theory of Games. Cambridge University Press, Cambridge.

Maynard Smith, J., Szathmáry, E., 1995. The Major Transitions in Evolution. Oxford University Press, Oxford.

Maynard Smith, J., Szathmáry, E., 1999. The Origins of Life: From the Birth of Life to the Origin of Language. Oxford University Press, Oxford.

McBrearty, S., Brooks, A.S., 2000. The revolution that wasn't: a new interpretation of the origin of modern human behavior. J. Hum. Evol. 39 (5), 453–563.

McHenry, H.M., 1994. Tempo and mode in human evolution. Proc. Natl. Acad. Sci. 91, 6780–6786.

McNabb, J., Binyon, F., Hazelwood, L., 2004. The large cutting tools from the South African Acheulean and the question of social traditions. Curr. Anthropol. 45 (5), 653–677.

McPherron, S.P., 2006. What typology can tell us about Acheulian handaxe production. In: Goren-Inbar, N., Sharon, G. (Eds.), Axe Age: Acheulian Tool-making from Quarry to Discard. Equinox Publishing Ltd, London, pp. 267–285.

Mednick, S.A., 1962. The associative basis of the creative process. Psychol. Rev. 220–232.

Miller, G., 2000. The Mating Mind: How Sexual Choice Shaped the Evolution of Human Nature. Heinemann, London.

Mitcham, C., Mackey, R. (Eds.), 1983. Philosophy and Technology: Readings in the Philosophical Problems of Technology. The Free Press, New York.

Mithen, S.J., 1990. Thoughtful Foragers: A Study of Prehistoric Decision Making. Cambridge University Press, Cambridge.

Mithen, S.J., 1996. The Prehistory of the Mind: The Cognitive Origins of Art, Religion and Science. Thames and Hudson Ltd, London.

Mithen, S.J., 2006. The Singing Neanderthals: The Origins of Music, Language, Mind and Body. Harvard University Press, Cambridge.

Mountcastle, V.B., 2005. The Sensory Hand: Neural Mechanisms of Somatic Sensation. Harvard University Press, Cambridge, Mass.

Napier, J., 1993. Hands. Revised by Russell H. Tuttle. Princeton University Press, Princeton, N.J.

Nitecki, M.H. (Ed.), 1990. Evolutionary Innovations. University of Chicago Press, Chicago.

Noad, M.J., Cato, D.H., Bryden, M.M., Jenner, M.-N., Jenner, K.C.S., 2000. Cultural revolution in whale songs. Nature 408, 537.

Noble, W., Davidson, I., 1996. Human Evolution, Language and Mind: A Psychological and Archaeological Inquiry. University of Cambridge Press, Cambridge.

Nowak, M.A., Tarnita, C.E., Wilson, E.O., 2010. The evolution of eusociality. Nature 466, 1057–1062.

Payne, R.S., McVay, S., 1971. Songs of humpback whales. Science 173, 585–597.

Pelegrin, J., 2009. Cognition and the emergence of language: a contribution from lithic technology. In: de Beaune, S.A., Coolidge, F.L., Wynn, T. (Eds.), Cognitive Archaeology and Human Evolution. Cambridge University Press, Cambridge, pp. 95–108.

Pitts, M., Roberts, M., 1998. Fairweather Eden: Life in Britain Half a Million Years Ago as Revealed by the Excavations at Boxgrove. Century, London.

Plummer, T.W., Ditchfield, P.W., Bishop, L.C., Kingston, J.D., Ferraro, J.V., Braun, D.R., Hertel, F., Potts, R., 2009. Oldest evidence of toolmaking hominins in a grassland-dominated ecosystem. PLoS One 4, e7199.

Renfrew, C., 1998. Mind and matter: cognitive archaeology and external symbolic storage. In: Renfrew, C., Scarre, C. (Eds.), Cognition and Material Culture: The Archaeology of Symbolic Storage. McDonald Institute, Cambridge, pp. 1–6.

Renfrew, C., 2004. Towards a theory of material engagement. In: DeMarrais, E., Gosden, C., Renfrew, C. (Eds.), Rethinking Materiality: The Engagement of Mind with the Material World. McDonald Institute, Cambridge, pp. 23–31.

Riel-Salvatore, J., 2010. A niche construction perspective on the Middle-Upper Paleolithic transition in Italy. J. Archaeol. Meth. Theor. 17, 323–355.

Riley, J.R., Greggers, U., Smith, A.D., Reynolds, D.R., Menzel, R., 2005. The flight paths of honeybees recruited by the waggle dance. Nature 435, 205–207.

Rothenberg, D., 2008. Thousand Mile Song: Whale Music in a Sea of Sound. Perseus Books, New York.

Schick, K., Toth, N., 1993. Making Silent Stones Speak: Human Evolution and the Dawn of Technology. Simon & Schuster, New York.

Schick, K., Toth, N., 2006. An overview of the Oldowan Industrial Complex: the sites and the nature of their evidence. In: Toth, N., Schick, K. (Eds.), The Oldowan: Case Studies into the Earliest Stone Age. Stone Age Press, Gosport, pp. 3–42.

Schlanger, N., 1996. Understanding Levallois: lithic technology and cognitive archaeology. Camb. Archaeol. J. 6, 231–254.

Seeley, T.D., 1985. Honey Bee Ecology: A Study of Adaptation in Social Life. Princeton University Press, Princeton.

Seeley, T.D., 1995. The Wisdom of the Hive: The Social Physiology of Honey Bee Colonies. Harvard University Press, Cambridge.

Seeley, T.D., 2010. Honeybee Democracy. Princeton University Press, Princeton.

Sinitsyn, A.A., 2002. Nizhnie kul'turnye sloi Kostenok 14 (Markina gora) (raskopki 1998–2001 gg.). In: Sinitsyn, A.A., Sergin, V.Ya., Hoffecker, J.F. (Eds.), Kostenki v Kontekste Paleolita Evrazii. Russian Academy of Sciences, St. Petersburg, pp. 219–236.

Striedter, G.F., 2005. Principles of Brain Evolution. Sinauer, Sunderland, Mass.

Swanson, L.W., 2003. Brain Architecture: Understanding the Basic Plan. Oxford University Press, Oxford.

Tattersall, I., 2009. Human origins: out of Africa. Proc. Natl. Acad. Sci. 106, 16018–16021.

Texier, P.-J., et al., 2010. A Howiesons Poort tradition of engraving ostrich eggshell containers dated to 60,000 years ago at Diepkloof Rock Shelter, South Africa. Proc. Natl. Acad. Sci. 107, 6180–6185.

Toth, N., 1985. The Oldowan reassessed: a close look at Early Stone Artifacts. J. Archaeol. Sci. 12, 101–120.

Toth, N., Schick, K., Savage-Rumbaugh, E.S., Sevick, R.A., Rumbaugh, D.M., 1993. Pan the toolmaker: investigations into the stone tool-making and tool-using capabilities of a bonobo (Pan paniscus). J. Archaeol. Sci. 20, 81–91.

Trivers, R.L., 1971. The evolution of reciprocal altruism. Q. Rev. Biol. 46, 35–57.

Vanhaeran, M., d'Errico, F., Stringer, C., James, S.L., Todd, J.A., Mienis, H.K., 2006. Middle Paleolithic shell beads in Israel and Algeria. Science 312, 1785–1788.

Villa, P., Soriano, S., Teyssandier, N., Wurz, S., 2010. The Howiesons Poort and MSA III at Klasies River main site, Cave 1A. J. Archaeol. Sci. 37, 630–655.

Wadley, L., 2001. What is cultural modernity? A general view and a South African perspective from Rose Cottage Cave. Camb. Archaeol. J. 1, 201–221.

Whitfield, C.W., et al., 2006. Thrice out of Africa: ancient and recent expansions of the honey bee, Apis mellifera. Science 314, 642–645.

Wilkins, J., Schoville, B., Brown, K., 2012. Functional analysis of ∼500 ka lithic points from Kathu Pan 1, South Africa. Paleoanthropology Society Annual Meeting, Memphis, Tennessee, USA.

Willoughby, P.R., 2007. The Evolution of Modern Humans in Africa: A Comprehensive Guide. AltaMira Press, Lanham.

Wynn, T.G., 1991. Tools, grammar, and the archaeology of cognition. Camb. Archaeol. J. 1 (2), 191–206.

Wynn, T.G., 1995. Handaxe Enigmas. World Archaeol. 27, 10–24.

Wynn, T.G., 1999. The evolution of tools and symbolic behavior. In: Lock, A., Peters, C.R. (Eds.), Handbook of Human Symbolic Evolution. Blackwell, Oxford, pp. 263–287.

Yellen, J.E., Brooks, A.S., Cornelissen, E., Mehlman, M.J., Stewart, K., 1995. A Middle Stone Age worked bone industry from Katanda, Upper Semliki Valley, Zaire. Science 268, 553–556.

Chapter 8

Climate, Creativity and Competition: Evaluating the Neanderthal 'glass ceiling.'

William Davies

Centre for the Archaeology of Human Origins, University of Southampton, Avenue Campus, Southampton, SO17 1BF, UK., swgd@soton.ac.uk

8.1. INTRODUCTION

Much recent attention has focussed on the perceived extinction event of Neanderthals, but little on how they changed and developed over the course of their long (200,000-year) existence. How did Neanderthals innovate and interact, and how did climate and competition impact on them? Those, of course, are large questions, and this paper will concentrate on the connections between them. It will focus on the Neanderthals of the Weichselian, looking at records from about 115,000 to 35,000 years ago, and evaluate the possible demographic structures of Neanderthal society. We cannot address the extinction of Neanderthals in isolation; indeed, we are very unlikely to be considering one single extinction, but rather a series of extinctions and replacements. Replacement need not involve *Homo sapiens*, except right at the end of Neanderthal existence.

A key debate in recent hunter–gatherer studies has been whether such peoples are passive respondents to their environments or active agents in making decisions irrespective of environmental conditions (e.g. Bender, 1985; Gamble, 1999, 2007; Gamble and Porr, 2005). Such debates are normally applied to modern hunter-gatherers, but they are important questions that should not be restricted to *H. sapiens*. Neanderthals, as large-brained hominins, must be allowed similar latitude: to what extent were their decisions dictated by the prevailing conditions of their environments? There is merit in adopting a broad demographic perspective, considering not just metapopulation estimates, but also the structure, constitution and interactions of groups and individuals. How sociable were the Neanderthals?

Discussions of creativity and transmission of ideas and innovations need to consider the demographic perspective, inferred from sites' sizes, layouts and behavioural signatures. Such comparisons crucially depend on a strong chronological framework. Patterns of extinction also need reliable dates, but extinction tends to be easier to determine if there is no subsequent replacement. We must consider, however, the likelihood that Neanderthal populations became locally extinct many times prior to their final disappearance. Such events would seem to fly in the face of any consideration that Neanderthals were active authors of their own destiny, independent of environmental fluctuations, but, of course, we should remember that active, independent decisions are not always successful.

Concepts of "refuge" and "demic expansion" are key to our understanding of demography and to behavioural change in relation to climatic and environmental fluctuation. However, such terms are seldom defined effectively by individual authors. Many archaeologists seem to assume that "refuges" are simply "asylums" or sanctuaries, to which populations oppressed by adverse climatic conditions elsewhere can retreat. Thus, hominins are allowed to remain the active authors of their own destiny, even if it means packing large populations into small "refugial" areas, and effectively depopulating the surrounding areas with minimal or no loss of human life. Such assumptions underpin the work of researchers such as Jochim (1987: 323) and others. The biological view of a refuge, as an area where individuals of a given species happen to (passively) survive in congenial conditions, while outside such environments, they become extinct (Dalén et al., 2007; Bennett and Provan, 2008) is something that is difficult for archaeologists to countenance. Surely large-brained hominins, such as Neanderthals, were sufficiently organised and proficient to be able to track habitats as they changed? If we are to demonstrate such habitat tracking by Neanderthals, then we also need to demonstrate that environmental change proceeded at a fairly gentle and regular pace across the landscape. Change whereby favoured habitats disappear from the landscape in a patchwork and mosaic fashion, leaving large (untraversable?) zones between scattered surviving patches ("refuges"), is not really considered. How far and fast do we expect hominins to be able to track changes in favoured habitat distributions: is there a rate at which their decisions are outpaced, and they are left isolated and prone to extinction, unless they can change their behavioural patterns? Such discussion depends largely on how receptive we think Neanderthals (and *H. sapiens*, for that matter) were to innovation and the adoption of new ideas. Extinction would be the reward of those too stubborn to change their *modus operandi*.

To set the argument of this paper in motion, we need to consider the following important themes and questions:

- How we reconstruct environments in the last glacial cycle experienced by Neanderthals: what scales and variables might have been important?
- How fast is "fast" when we consider environmental and climatic change, and what spatial scales of landscape occupation should we model for Neanderthals?
- In cycles of "refuge" and "expansion", mobility is a key consideration. How much variation should we expect between restricted and expanded mobilities, respectively, and do they necessarily correspond to high and low population densities, respectively?
- Differential levels of mobility (whether climatically or socially driven – or both) were surely crucial in affecting the scales and rates of generation and transmission of new ideas. Should we therefore assume greater levels of generation and transmission of such ideas when population density and mobility are relatively restricted, or should we assert that increased mobility between more scattered populations is equally, if not more, effective at generating and spreading innovations? Certainly more mobile, networked, populations would have the potential to transmit ideas over further distances.

Such questions, if not answering the question of the final extinction of Neanderthals, will serve to set it into a broader, more interesting framework. We need to neutralise the sense of inevitability about Neanderthal extinction; Neanderthals seemed to attain the peak of their success in the period between 115,000 and 35,000 BP; so a broader view is essential. We now stand at the threshold of a new chronological revolution, with the alluring possibility of being able to obtain new, more reliable dates for Neanderthal sites, especially the last ones (after 55,000 BP) (Higham, 2011; Higham et al., 2010; etc.). Dates by themselves, though, will not answer our questions. We need to have clear heuristic frameworks by which we can interpret them and choose our scales of spatio-temporal analysis.

8.2. POSSIBLE CAUSES AND SCALES OF CHANGE

The term "innovation" needs definition for the purposes of discussion. One possible definition is that it is the generation of unprecedented artefacts or behaviour within a given group. However, there are problems with such a definition at the *human scale*: what counts as "unprecedented"? For example, if a behaviour were intermittently reinvented within a society, but with long gaps in between the re-manifestations – such that its knowledge were lost to living memory (perhaps three to five generations after the abandonment of an innovation) – could it have been seen (repeatedly) by its discoverers as unprecedented? Innovation, as has been pointed out by many researchers, requires not only the generation of new ideas, techniques and artefacts, but also the *recognition* of their novelty by their inventors, and thus their subsequent recreation and development. If an innovation is reinvented after a period long enough for its previous loss to be entirely forgotten, it would effectively be seen by that society as something novel. In such situations, it would be rather pedantic of us to consider them reinventions. Thus, I do not propose to separate reinventions from innovations *sensu stricto*, especially given the relatively coarse spatio-temporal resolutions of archaeological contexts.

We must also recognise that innovation is not directly equivalent to social and technological progress. Removing inevitability from innovation is an important step. Information and ideas can be lost in societies (e.g., the loss of skills in making bows and arrows, kayaks and snow houses among the Polar Inuit of Anoritoq, Greenland, owing to the deaths of many adult males from an introduced plague (Rowley, 1985)), and it is misleading and erroneous to suppose that human development consists of the accumulation of an increasing array of beneficial and sophisticated inventions – none of which can or would be discarded, owing to their self-evident indispensability to their "host" societies. Such hindsight corrupts the archaeological debate, particularly with respect to "behavioural modernity". Behaviourally "modern" traits were not necessarily self-evidently essential to their practising communities, to be retained at all costs: some may have been deemed expendable or unnecessary in certain circumstances, or even lost through behavioural drift or group extinction.

Such a debate cuts to the heart of our theories of evolution, and ultimately derives from the conflict between Lamarckian and neo-Darwinian views of evolutionary process. These two evolutionary perspectives make important, but different, assumptions about how innovations originated, developed and spread. Lamarckism assumes that necessity is the mother of invention, and that past peoples must have responded to such (environmental) challenges by innovating and adapting their behaviour accordingly. In general, Lamarckian typological systems favour the identification of ever-increasing complexity. They can be characterised as "evolution to order"; organisms have to either enlarge or develop characteristics or organs they hitherto seldom used, or else they make use of ones they have spontaneously generated to fulfil a need by "efforts of… inner feeling" ("par des efforts de… sentiment intérieur":

Lamarck, 1809: 234). In contrast, neo-Darwinian evolution emphasises a law of divergence (radiating and branching), rather than strict convergence: variability occurs through descent, and is selected for by environmental conditions. It might be argued that Lamarckism is more amenable to the recognition of active development and innovation of cultural traits than neo-Darwinism, but in truth, any adaptationist explanation (including Lamarckian ones) will always have difficulty in explaining behaviours that transcend in complexity and variation what should be expected from environmental parameters. In other words, Lamarckism is no better than neo-Darwinism at explaining why we see the cultural variation we do in the Palaeolithic record (see end of this section).

In the Palaeolithic, discussion has tended to focus on whether adverse or beneficial environmental conditions most favour innovation: do "good times" allow more leeway for experimentation, or is there nothing to lose in "bad times"? However, I here propose an alternative. Foley's (1994) analyses implied that climate change had little effect on hominin speciation, but rather more on species extinction. Extrapolating this idea to the effects of climate change on cultural variation, we can propose that innovations were fairly constantly generated in the Palaeolithic, but selection pressures were more severe against them in periods and places of (more) adverse conditions. Climate might therefore have had little effect on the rate of innovation, but rather more on whether those innovations became extinct or prospered. This approach – more neo-Darwinian than Lamarckian – allows us to avoid the problems of attributing positive or negative environmental effects directly to the speed and iteration/tempo of innovations.

Differential scales of change are essential in testing notions of innovation in the archaeological record. The easiest subdivision is a tripartite one:

1. **Large scale:** change from one technocomplex or archaeological "culture" (*sensu* Clarke, 1968) to another (*suite* change).
2. **Medium scale:** change from one tool-type to another (*element* change).
3. **Small scale:** refinement and/or change in the form of a particular tool type, perhaps reflecting changes in the construction/assembly of composite tools (*morphological* change).

Taken in combination, these three scales account for the changes seen in archaeological successions and assemblages. Reliable chronologies not only serve to establish diachronic or synchronic change at these three scales over space, but also have the potential to tell us about the speed and frequency (iteration) of innovation. Do innovations always persist over time and across space? Or do they die out, and sometimes get reinvented independently? Nowell and White (2010: 73) have recently distinguished between innovation (creation of totally novel traits) and "inventiveness" (reworking/adapting a trait that already exists). The question of how we distinguish what is inventive from that which is innovative in archaeology can be answered with regard to the tripartite scheme outlined above. Inventiveness is likelier to be identified at the small scale, and to a rather lesser extent at the medium scale. Innovation is much clearer to identify at the large and medium scales.

The above-mentioned three scales of archaeological change also imply different scales and tempi of variation, with the small-scale one happening the fastest (a fraction of someone's lifetime) and large-scale change happening most slowly (perhaps spanning several generations). Neo-Darwinian evolutionary perspectives would favour behavioural "descent with modification" explanations of novel traits, with large-scale change being driven mostly by accumulated smaller-scale changes, though perhaps occasional replacement by intrusive suites that had developed elsewhere and subsequently spread into the study region (effectively a cultural form of allopatric speciation (Mayr, 1970: 279)) might be considered. Lamarckian views, on the other hand, would tend to emphasise the larger scales of innovation and their relatively sudden appearance in the record (whether autochthonous, or dispersed into the study region from elsewhere).

Much attention has been devoted to "transitional" archaeological industries, particularly at the end of the Neanderthals' existence (e.g., d'Errico et al., 1998; Allsworth-Jones, 1986, 1990). However, the rationale for seeing something as "transitional" seems often to derive from the identification of "mixed" characteristics. These characteristics can encompass all scales of change, from small (primarily technological analyses) to medium (typological) and large (cultural/cognitive) ones. Two main assumptions are made in such studies of "transitions":

1. We can identify cultural change as if we were insiders of the cultural group (the "emic" view: Harris, 1979: 32) and
2. That change moves in discrete, identifiable steps, with each step representing an essentialist phase (Henneberg, 2009).

The first assumption, of course, is deeply questionable, especially given the incomplete material records we have from the Palaeolithic. Nevertheless, researchers studying the Palaeolithic have been identifying supposed ethnicity from tool types and technological traits, and inferring function from intuitive readings (and, later, use-wear and ethnographic studies) of tool forms for well over 150 years (e.g., Lubbock, 1864; Lartet and Christy, 1864, 1875; Breuil, 1912; Bordes, 1961; Semenov 1964; d'Errico et al. 1998; Vanhaeren and d'Errico, 2006). In the last fifty years, the concept of the *chaîne opératoire* (Leroi-Gourhan, 1964) has risen to increased prominence, seeking to characterise and quantify the "operational chain" of decisions and choices made at every stage of manufacturing and discarding artefacts (e.g., Geneste, 1991; Chauvière, 2005; Faivre, 2006; and references therein). However, such approaches still tend to assume that Palaeolithic groups used only a very limited range of inherited behaviours – an assumption falsified by Neanderthal sites such as Maastricht-Belvédère (de Loecker, 1994; Roebroeks et al., 1997).

The implication of sites such as Maastricht-Belvédère is that hominins were drawing on a repertoire of different "operational chains", rather than following just one per raw material type or class. Thus, our hopes of being able to recreate the thought processes and ethnic markers of different past hominin groups rest on shaky foundations, and we cannot be certain that changes in *chaîne opératoire* through a given site sequence or for an archaeological industry necessarily reflect cultural change as perceived by the original makers.

The second assumption is closely related to the first one: based on taxonomic ideas that extend back beyond Linnaeus, we often perceive change as moving in a stepwise fashion, as a series of progressive jumps condensed into discrete steps, with intermediate static or stable phases. Anything that does not contribute to the general progress (to the next step) is categorised as "noise," or irrelevant variation. "Transitions" are thus perceived as directional, "progressive," processes, moving from one suite of characteristics to another. The "transitional industries" (Châtelperronian, Uluzzian, Szeletian, etc.) of Europe (c.45,000–35,000 BP) are widely assumed to display a progression from Middle to Upper Palaeolithic technologies and *chaînes opératoires*, and are also simultaneously viewed as evolutionary "dead-ends," becoming extinct with their presumed Neanderthal authors (cf. d'Errico et al., 1998; Kuhn, 2003). Only authors such as d'Errico et al. (1998) have argued for acculturation of *H. sapiens* by autochthonous Neanderthal groups, thus ensuring survival of Neanderthal ideas beyond their demic persistence.

All classificatory systems generalise, to varying extents, from the particular. Henneberg (2009) has distinguished between "essentialism", which asserts the binary existence of taxa (extant versus extinct), and neo-Darwinian views of individual variation. In essentialist approaches, individual specimens are assumed to be representative of the wider species, and variation across time (and space) is compressed into a generalised set of traits. Change between one taxon and another, according to Henneberg, in an essentialist model is assumed to be catastrophic (sudden replacement of one set of traits with another). Neo-Darwinian studies of individual variation seek to establish points of similarity or difference between contemporary individuals, rather than describe the similarity of individuals to a single, immutable type (Henneberg, 2009: 69). Applying this distinction of approaches to archaeological taxonomy, it would appear that analyses that emphasise "transitions" fit more into an essentialist mode than one of neo-Darwinian variability. Whilst approaches that emphasise similarities and differences between contemporary archaeological assemblages are also well established in archaeology, I cannot think of any that avoids the use of "type sites" or "type assemblages" as benchmarks to calibrate similarities/differences. That is partly because many identifications of archaeological industries occurred in the latter half of the nineteenth century, and in the first half of the twentieth one, when absolute dating was scarcely possible, and the use of "type sites" with long stratigraphic sequences became essential in constructing frameworks of reference for analysing archaeological material. Similar, often single-occurrence, archaeological assemblages or sites could be compared (and thus relatively dated) against the type assemblage. However, we now have new ways of obtaining absolute chronologies for elements within assemblages, and so we can start to advance beyond simple reliance on a limited number of "type assemblages". We can now begin to compare penecontemporary variation in form and technique between artefacts from different assemblages. Many of these advances are part of progress in radiocarbon dating (Bronk Ramsey et al., 2004), affecting the last twenty millennia or so of Neanderthal existence. It should be hoped that similar advances can ultimately be made for earlier phases of Neanderthal existence (see below).

It seems to have become axiomatic among many researchers that while hominin physical evolution is governed by neo-Darwinian processes, Lamarckian ones better account for the development and transmission of culture (e.g., Zilhão, 2006: 65; Gould, 1980: 71; Spuhler, 1984: 13, cited in Cullen, 2000: 58). However, this paper will argue that this assertion rests upon ill-defined foundations. While Lamarckism may ostensibly provide a process of "cultural" transmission of (nongenetic) acquired characteristics, it makes major assumptions that have tended to fetter our analyses rather than explain them. Lamarckism's notable assumptions are:

- That we can identify discrete types and cultural units that can be transmitted (the typological approach). This is an essentialist assumption, which requires rigorous testing (direct dating, artefact morphometry, experimental archaeology, etc.).
- That periods of change should be driven by "need", and be short-lived. This determinist view is also partly informed by essentialism, and could help to explain why there is so much preoccupation with "transitions", "crises" and "revolutions" in the archaeological literature.
- That no "leap-frogging" over developmental stages can happen. Evolution proceeds in a linear, directional

manner, normally propelling cultural systems towards ever-increasing complexity or perfection (or both): i.e. "progress". Sometimes reversals occur ("devolution" or degeneration), as do small side branches, but such examples serve to demonstrate the overall validity of progress.

- The environment *determines* rather than selects change; it directs the *needs* of organisms, which in turn affect their behaviour or body shape. Change is "rational" or "objective", and transformation is achieved through use or disuse.

It is thus not helpful to assume that only Lamarckism can deal with "cultural" issues. Although genetics were incorporated into Darwinism to form the neo-Darwinian synthesis, it does not follow that neo-Darwinian evolution has no explanatory mechanism for the transmission of nongenetic behavioural variables. Nevertheless, relating neo-Darwinian selectionism to "hot" (actively respondent) societies (in addition to those perceived as "cold", or passive, respondents to environmental change) (Bender, 1985) requires careful consideration. These points will be elaborated later in the discussion.

Amalgamating the above themes, we can start to analyse change in the Palaeolithic record in new and productive ways. We need to consider change at different material culture scales (from small to large), and in terms of its structure and (social) organisation. Nowell and White's (2010) definition of inventiveness allows evaluation of small-scale change, measuring the morphological modification or recombination of individual elements. However, such reorganisation of deckchairs may not be enough to avert a titanic social disaster, and larger-scale change (innovation) will be required. Innovation requires that new elements are added to the mix, and perhaps others taken away. The big question is whether inventiveness and innovations are responsive to climatic and social stimuli/dictates, or whether they proceed independently, and are then selected for or against by such conditions (cf. Foley, 1994). Essentialist/Lamarckian or neo-Darwinian approaches determine how we view variation in the archaeological record: the former would see "directionless" small-scale variability as "noise", with transitions occurring in (catastrophic) steps, while the latter conceives "transition" as a constant process, subject to intermittent selection. "Noise," in a neo-Darwinian view, could only be seen in small-scale attributes that were selectively neutral; otherwise, change should be seen as responsive to environmental (or social) requirements.

8.3. SOCIAL PROCESSES OF POPULATION EXPANSION AND CONTRACTION

Another dimension of archaeological change is the demic aspect, related to different scales of population change and movement. Essentialist schools of thought would tend to attribute major "transitions" to acculturation, sudden changes of population or social structure/organisation, or a combination of all three. Here, acculturation is simply defined as the social transmission ("diffusion") of ideas from one population unit to another. As with archaeological culture, these (Neanderthal) population units would have existed at different scales, from the metapopulation at the largest scale, narrowing to the breeding group, the family and the individual.

8.3.1. Innovation in Mobile Groups

It is one thing to assume acculturation between a set of sedentary, territorial hominin groups. However, this scenario is unlikely to have pertained for many periods and regions in the Palaeolithic, for climatic oscillations would have affected the distributions of resources over time and space. Instead, we have to consider spatio-temporal variations in population densities; people were not distributed evenly across the landscape, reflecting the availability of resources. The abundance and predictability of resources are key in determining the levels of mobility of hunter-gatherers (e.g., Dyson-Hudson and Smith, 1978), which in turn affects their spatio-temporal distributions. Neanderthals did not occupy the whole of their maximal range throughout their existence; large areas were unoccupied (perhaps never occupied), while others saw consistent levels of occupation, e.g., SW France. If we were to divide Europe, and western and central Asia into a grid, some grid cells were never occupied, others were inhabited intermittently, and the remainder saw consistent occupation. Thus, the Neanderthal world was a patchwork of occupied and unoccupied cells/areas at any given time of their existence. Such irregular patterning across the landscape would have had repercussions for levels and scales of connection (and thus intergroup acculturation), although such degrees of connectedness might be more difficult to predict if we factor in mobility of one group to an area close to or within the area of another one, traversing an unoccupied zone Fig. 8.1.

Restoring a sense of mobility to Palaeolithic hunter-gatherers is relatively easy, but our bigger challenge is to try to reconstruct pathways through the landscape, which might help us to connect "patches" of penecontemporaneous occupation. In other words, we are trying to grasp the *structure* and *directionality* of prehistoric movements. Gamble (1999, 2007) has sought to address this key issue of pathways; three main lines of evidence can be used to test such ideas:

1. The distribution and density of contemporary archaeological material in the landscape (Isaac's (1981) scatters and patches model);

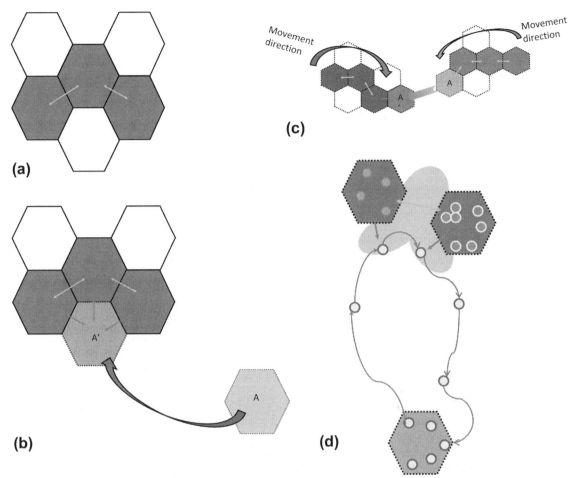

FIGURE 8.1 Mobility and the potential for transmission of ideas. (Key: White cells = unoccupied; solid lines = social territorial boundaries; dotted lines = open social range boundaries; yellow arrows = interactions between established groups; red arrows = socially restricted interactions between newly adjacent groups.) **(a) Sedentary, packed populations**. Regular, consistent transmission of new ideas, over extended periods, is expected between closely packed, established groups with low/restricted mobility (see Table 8.1). Unoccupied areas form barriers to transmission, with little or no spread of new ideas and traits across them. The mobility and transmission systems are essentially *stable* over long periods, with consistent levels, rates, scales and network types of information exchange. **(b) A mobile group (A) moves into proximity (A′) with sedentary, packed populations**. Regular, consistent transmission of new ideas, over extended periods, is expected between closely packed, established groups with low/restricted mobility (see Table 8.1). Unlike Fig. 8.1a, however, the immigration of an allochthonous group (A) into an unoccupied area (A′) adjoining these established groups creates potential for the transmission of new ideas, at varying levels and rates, and at different scales of social network intimacy (perhaps operating at the effective network scale, but perhaps more probably at the extended and global scales (*sensu* Gamble, 1999)). The intensity of information exchange might be related to duration of stay of the group in A′, with potentially more (and a wider range) of information being exchanged with increased duration of stay. Mobility is mostly localised and constrained, with the single high mobility of one group being the exception (see Table 8.1). **(c) Two string of pearls dispersing populations are connected by limited saltation movement**. Consistent, though potentially irregular, transmission of new ideas between dispersing and fissioning groups over extended periods. Any spatio-temporal irregularities in transmission would be expected to be caused by the need to balance resource gathering requirements against the maintenance of social connections in high mobility contexts. Down-the-line movement of people and ideas should be expected to occur, reflected in transmission of novel traits along extended chains of occupation ranges. There is some potential for short- or long-term rapid transmission of innovations over long distances, depending on the scale and frequency of saltation (instantaneous "leaps" of a group from one area to a more distant one), indicated in this diagram by the movement of a group from A to A′ (thus connecting it to a string of pearls dispersal moving in the opposite direction). Mobility is generally high, though incremental and stepwise, and social boundaries are open and permeable, facilitating fluidity in exchange of ideas (from intimate to extended network scales) between mobile groups. There are less frequent opportunities for exchange of ideas through saltation events, though when these do occur, they may vary in their intensity (intimate to global network scales) by the degree to which the newly adjoining groups are open in their social networks (see Table 8.1). **(d) Short-term sequential connection of a highly mobile group with two other (non-packed) groups, through saltational extension of meta-range activities.** (Grey hatching denotes interaction zones, where interaction can take place.) Groups are expected to exhibit long-term high mobility (see Table 8.1), at varying spatio-temporal scales; there are no firm, adjoining social boundaries, packing, or directional down-the-line transmission of ideas. All three groups have open social network boundaries and rely on mobility, which bring them into contact (either synchronously or diachronically) permitting the transfer of ideas between them, at varying network scales (possibly including Intimate scales, though more likely to operate at effective to global scales, owing to short-term nature of contacts).

2. The movement/circulation of raw materials around the landscape from source to final place of discard; and
3. The modifications that such raw materials undergo as they are selected, used and possibly reworked across the landscape ("artefact biographies").

If hominin actions are orientated along these pathways (exemplified in the acquisition and discard of resources), then it must be admitted that some potential pathways seem more obvious to us than others, e.g., river valleys. Less obvious pathways – while we know they must have existed – are more difficult to identify if we have no obvious topographic features with which to connect them. However, the careful use of lithic raw material provenance data can yield useful insights into areas that were exploited by Neanderthals, where sources can be identified, and connected to the final discard of artefacts in archaeological assemblages, as well as related to the hunting and butchery patterns of faunal resources (Raynal et al., 2007). It is even possible to identify regions that Neanderthals did not seem to visit, at least to gather lithic raw materials (Raynal, pers. comm.). It seems reasonable to assume that lithic raw materials were collected during hunting and gathering trips (Binford's (1979) "embedded procurement", which is derived from modern hunter–gatherer behaviour at high latitudes). If we can talk of "embedded procurement", it also seems sensible to speak of "embedded discard" as well, when artefacts are discarded in connection with other (related) economic and social activities (see Gamble, 1999: 358–361). Thus, the density and context of the contents of archaeological assemblages need to be considered in tandem.

Innovation should be expected to occur in any given social or economic activity location, but we need clear spatio-temporal models in order to plot its distribution and spread. The directionality of human movement should have an effect on the generation and spread of novel traits. In addition, models of demic dispersal into unoccupied territory should be considered for the spread of innovation within expanding populations – a process that may or may not bring them eventually into contact with other, expanding or sedentary groups. Two resource-related models of hominin directional dispersal have been proposed: the "string of pearls" and "leapfrog"/saltation models (Anderson and Gillam, 2000). The string of pearls model envisages a series of contiguous territories, forming chains of connected occupation regions; such a spatial organisation might lead us to expect "down-the-line" transmission of novel traits, from one territory to its neighbour, perhaps accelerated if a group is displaced from its territory and forced to move on (Fig. 8.1c; Table 8.1). Changes in form (the lowest scale of change) would be expected to show subtle changes with increasing spatio-temporal distance from a source, reflecting variations deriving from down-the-line interactions and modification. A saltation model might show "leaps" of novel traits across considerable distances of unoccupied landscape, from one territory to another, within short time spans, assuming that the distance is traversable (if it is not, then presumably novel traits will be restricted to their territory of origin). If the traits do move beyond their point of origin, then there are several possible transmission methods: (a) through kinship ties between different territories, (b) through connecting meta-territorial movements in the "interstitial" zones between separate territories or (c) displacement of a group from its territory to a suitable area close to other separated territories (Fig. 8.1; Table 8.1). Here, (b) and (c) are assumed to occur independently of kinship connections, making their dissemination of novel traits essentially stochastic; however, it is also entirely possible that they could be linked to kinship, and thus not too dissimilar in drivers or outcomes from (a).

Tracing the spread of innovations across a spatio-temporal grid is, of course, not a new approach. The work of Torsten Hägerstrand (e.g., 1967a, 1967b) is of great relevance to this discussion. Hägerstrand (1967a, 1967b) assumed that novel traits would spread largely via face-to-face transmission; such a model would fit well with Gamble's (1998) intimate and effective networks, where physical copresence was essential to the exchange of information in Neanderthal society. However, individual and group residential mobility do not seem to be an intrinsic part of Hägerstrand's model; instead, the underlying assumption is of essentially sedentary (agricultural) societies, though with some mobility between settlements (Hägerstrand, 1967a: 8). Although "people are constantly in motion" (Hägerstrand, 1967a: 8), with every individual possessing a unique field of movement, their mobility is essentially centred, radiating outwards from a (home) centre. From these individually centred patterns of movement, Hägerstrand extrapolates that:

"[t]he talking and listening individual is part of a huge, world-embracing network of links. A good many observations suggest that this network has a definite spatial structure which probably is rather stable…" (1967b: 7).

Hägerstrand (1967a, 1967b) thus sought to explain his nebular spatial distributions – with densely populated cores and decreasingly populated peripheries with increasing distance from the centre – using a cellular model. Innovations would occur within certain cells, and then be transmitted (by human agency along networks) between those cells and primarily adjoining ones. Some allowance was made for rapid long-distance transmission of traits, depending on whether individuals (or groups) operated at the local, regional or international scales (or combinations of some or all of those scales) (Hägerstrand, 1967b: 8).

TABLE 8.1 Future Testing of the Potential Effects of Different Scales of Hominin Mobility on the Transmission of Ideas (cf. Binford, 1980; Bettinger, 1991)

	Localised	High Mobility
i) Raw material use	Local raw materials preferentially used. "Exotic" raw materials scarcely used; only available through exchange with other (neighbouring groups). **Thus, depending on the quality and variety of local raw materials, innovations are likely to be visible across the whole assemblage (generalised/nonspecific innovation and inventiveness).**	Varying proportions of "exotic" raw materials apparent, depending on scales of movement, discard patterning and levels of tool/blank/core *curation*. There is some potential for *extended* distance if direct exchange with widely spaced, distant were practised. **Thus, more specialised innovation/inventiveness is evident, linked to activity specialisation throughout the occupied range: *specific* uses of particular raw materials.**
ii) Technology	Assemblages are expected to show a consistent diversity of tool types, with very little spatial distinctiveness or structure seen in assemblage patterning throughout the intensively occupied range. **Thus, there is much "noise" in the patterning, as this lack of spatial separation in site function affects our ability to detect change. Unless there is extensive contact with neighbouring groups, any innovation/inventiveness will tend to be locally restricted, and at strong risk of extinction through cultural drift. Any *transmitted* novel traits will change rapidly across space (and time) as they move between closely spaced groups.**	Spatial (inter-site) structure in the distribution of assemblage types is expected to be easier to detect, with the patterning being more widely spaced and often more functionally discrete. As groups range more widely, they can transmit novel traits over longer distances with less evident loss of information or modification of design as ideas are exchanged. **Thus, innovations have much more potential for distant dispersal (and thus overall survival?), with less chance for incremental information loss and modification through down-the-line transmission — even if some groups suffer extinction.**
iii) Subsistence	Exploitation of a wide range of food resources is evident, with spreading of risk across a relatively broad spectrum of species. This strategy is dependent on enough resources being found year-round to sustain localised occupation. Groups would therefore be expected to locate themselves on or near ecotones to provide suitable resource security; physical storage of surplus resources, *if present*, would be restricted primarily to residential sites. **Thus, because subsistence activities are focussed at the local scale, any innovations in resource exploitation will be relatively "noisy" and stable; innovations will only become more strongly selected for if requirements for increased environmental productivity demand intensification of subsistence methods. Innovations in localised societies might be expected to be spatially restricted (with variable potential for spreading far afield), though intensively present in the landscape (creating the "noisy" innovation signals).**	Subsistence activities are focussed on a restricted range of species, perhaps also indicating specialisation in resource selection, processing and consumption strategies. High mobility designed to exploit patchy/clumped food resources that require directional movements between them. Physical storage of (surplus) food resources spread throughout landscape, close to their places of processing. Transport costs are relatively high for parties to move resources to the rest of the group, unless *whole group* is highly mobile and participates in resource gathering. **Thus, any subsistence innovations will be concentrated, though clumped, in the landscape. The demands of responding to scattered, often highly mobile, species will potentially ensure that innovations are transmitted over long distances, and geared to increasing subsistence security in unpredictable situations. If whole group (residential) movement occurs over long distances between resource patches, we should expect: innovations in physical storage systems geared towards portability (e.g., drying), and the wide-ranging exchange of new subsistence ideas when two different groups meet. If specialist parties move over long distances, we should expect: innovations in physical storage systems geared towards large-scale transport and câching in the landscape, and the specialised exchange of new subsistence ideas when two different parties meet from separate groups.**

TABLE 8.1 Future Testing of the Potential Effects of Different Scales of Hominin Mobility on the Transmission of Ideas (cf. Binford, 1980; Bettinger, 1991)—cont'd

	Localised	High Mobility
iv) Site variability	A general lack of inter- and intrasite differentiation should be expected, partly owing to the palimpsest overlap of different occupations at the same locations. Generalised, rather than function-specific, (residential) sites are common. **Thus, a lack of intra- and intersite structure will affect how innovations are generated and transmitted (potentially to anyone in proximity to the innovator), leading to an unstructured generation and transmission of new ideas. Very "noisy" signal, perhaps concentrated round hearths and other clear foci in the site, especially when examined at the landscape scale: blurring of social boundaries between generalised and specialised innovations?**	A clear patterning of social space (at intersite, and perhaps intrasite, scales) should be expected, though consistent patches of resources might see more palimpsest overlap of activities. A variety of site types is common: residential and function-specific. **Thus, inter-site structure will primarily determine the type and success of individual innovations and their transmission. Some potential for blurring of generalised and specialised innovations, particularly in residential sites, but clearer evidence of innovation in more specialised activity sites and/or intrasite specialist areas. Therefore, the scale, type and transmission of innovation depends greatly on whether the whole group is highly mobile or whether a specialist part of it is, which will affect the potential scale of transmission of new ideas.**
v) Social/symbolic behaviour	Spatially defined symbolic activity is evident, evenly spread across the landscape. Clear boundaries are expected for symbolic expression, reflecting the need for contiguous populations to express social similarities and differences. Burial evidence might be expected to be common, and to show diversity of practice and consistently high spatial densities, as part of groups' connections to their territories. **Thus, symbolic and social innovations will be confined to particular territories (reflecting spatially limited networks), often being used to reinforce boundary differences. Innovations may never pass beyond those boundaries; if they manage to do so, they may undergo significant change with increasing distance from the point of origin, as replication variation and modification affect the manufacturing techniques, final form and (potentially) function (owing to lack of face-to-face contact with the originators of an idea).**	Much more spatial variation in the scales of symbolic systems is evident, though consistently extending for longer distances than seen in localised social systems. There is strong potential for more focussed exchange of particular artefacts at long-distance scales, with symbolic and economic activity being thinly spread or "clumped" in the landscape. Social networks open, with weak territorial boundaries. Burials are uncommon, and irregularly distributed in the landscape; consistency of mortuary practice expected over large areas. **Thus, symbolic innovations will have the potential to spread directly (rather than across extended chains of social connection) over long distances. If residential movements are extensive, the peripheries of ranges will show initial cross-fertilisation of ideas with neighbouring groups, which will then be carried back to the core area. If residential movements are more restricted to the core areas of the range, then any exchange of ideas in the peripheral zones is more likely to be restricted (certain classes of symbolic artefact and tools closely connected with special activities).**

Hägerstrand's ideas provide a useful starting point for archaeologists wishing to plot the origination and transmission of behavioural traits in time and space, but there are problems for researchers studying the Palaeolithic. Certainly, Neanderthals are often assumed to have existed at low population densities, though they may have been more "tethered" to established territories and concentrated in the landscape than *H. sapiens*, tending to move over relatively restricted distances in smaller groups (e.g., Lieberman and Shea, 1994; Raichlen et al., 2011; Finlayson, 2004: 115–117; Verpoorte, 2007; White and Pettitt, 2011). However, we need to be careful in how we apply concepts and models based on Hägerstrand, and need to ask questions such as:

1. If Gamble (1998) is right, then face-to-face communication (as modelled by Hägerstrand) should be expected to predominate in Neanderthals. However, the cumulative effects of such communication should result in down-the-line transmission (perhaps creating descent with modification): what are the social requirements for such transmission? What social network structures should we model, and what might their effects on transmission be?

2. Why would systems necessarily be stable? Any stability would presumably depend on consistency in the availability and predictability of resources; however, the climatic and environmental structure of the late Pleistocene is known to have fluctuated, often within hominin lifetimes. What were the social and economic effects of such fluctuations: survival in some places (ensuring some degree of stability of transmission?) and extinction in others (resulting in localised losses of innovations)? "Active" models of refuges (e.g. Jochim, 1983, 1987) might imply that Neanderthals might migrate into refuges from adjoining regions, thus altering the balance and intensity of transmitted traits between individuals and groups (see below).

3. What are the effects of long-distance mobility on transmission systems? In effect, what effect would high-mobility have on our modelled spread of an innovation from one cell to an adjoining one? Some cells would be entirely bypassed, leading to "leap-frogging" of that innovation across space. In addition, movement of an individual or a group from one range to (close to) that of another would effectively serve to contract spatial gaps between groups, allowing innovations to "arc" between two normally widely spaced groups in a very short time (Fig. 8.1b–d).

Other *modes* of analysing the transmission of traits have arisen more recently. A notable example is the cultural virus model of Cullen (2000). This model explores the transmission of innovations to some extent independently of direct human agency: cultural viruses are ideas that have an existence independent of the body, to varying degrees (though still dependent on lines of transmission). They are transmitted through socially mediated learning, and their application and development might confer increased reproductive success on individuals using them. In other words, this model is classically neo-Darwinian. Cultural viruses can conceivably effect change at every scale – suite, element and morphological (see above) – and thus can operate at all social spatio-temporal scales, from the momentary to transgenerational. Their reproduction is not guaranteed by the reproduction of their "hosts", or vice versa, and thus they can survive and develop irrespective of the survival of innovators or transmitters if social connectivity occurs regularly and covers a wide range of behaviours. Cullen (2000: 176–177) emphasises that cultural phenomena benefit passively from "unconscious *and conscious* attempts by human individuals to control their lives", with hominin "intentions" and "emotions" being more prominent as selective pressures than "unconscious psychological factors," as in Dawkins's memes (p. 176). According to the cultural virus theory, ideas and artefacts require external agencies, such as hominin minds, to reproduce and transmit themselves within their environment. Hominins will be, at times, aware of their reproduction of cultural traits, and often why they are doing so (Cullen, 2000: 177). Human intentions, purposes and planning thus become "fundamental elements of the psychological landscape – the idea's 'environment' – to which populations of ideas adapt" (p. 177). Individual cultural viruses (ideas) can vary greatly in size and complexity, with comprehensibility varying according to the abilities and relationship between teacher and learner. Face-to-face transmission is an important part of the cultural virus theory.

Before we progress to assessing Neanderthal innovation and creativity, the above discussion needs to be summarised and key points for testing emphasised.

1. Change can occur at different social and spatio-temporal scales: small (morphological), medium (element) and large (suite) scales. Small-scale changes should have the highest turnover (perhaps even days or weeks) and be restricted to the intrasite scale, while large-scale change should occur over longer periods, ranging from decadal (intergenerational?) to centennial (perhaps even millennial) spread across wide areas.
2. Creativity should be seen as a constant process (neo-Darwinian approach), generating variation from which certain novel aspects are selected, reproduced and developed. Innovative and inventive traits can persist (at scales outlined in (1) above) or become extinct.
3. Demography (metapopulation scale, spatio-temporal population density and social structure) needs to be considered in the development and transmission of novel traits. However, (effective) population size alone is an inadequate explanation for innovation (e.g., Shennan, 2000, 2001); larger populations may increase the number and rate of invention, but social structure and connectedness is a better explanation for transmission of new ideas, especially in mobile populations.
4. The importance of different scales of mobility needs to be recognised in the generation and transmission of novel traits. More sedentary populations will transmit (or not) their ideas in different ways, and perhaps at different rates, from their more mobile contemporaries/neighbours. Do innovations spread through direct face-to-face contact or can they spread far, over chains of social connections ("descent with modification" across attenuated sequences of face-to-face contact) (Fig. 8.1)?
5. To what extent is their existence *independent* of the populations that invented them? Does extinction of a population and its innovations happen at the same time, or at different times? If such cultural viruses can persist independently of the fate of their progenitor groups, do we need different reasons (from those ultimately based on variants of face-to-face transmission) to explain their extinction?

8.4. PRECISION AND ACCURACY IN DATING THE SPREAD OF BEHAVIOURAL NOVELTY

Key to the testing of the above key points is reliable chronometry, whose accuracy and precision allow us to plot change in time and space, and also to assess the impact of contemporary climatic conditions and fluctuations. If we adopt a neo-Darwinian approach, we should expect variation and generated at a more or less constant rate, with selection operating on that variability particularly at times of greater social stress. If innovations can be demonstrated to have occurred independently of climatic or environmental stimuli, then other reasons must be found to explain them: perhaps greater efficiency or a social preference for new forms.

Unfortunately, at present our chronology for Neanderthal existence is patchy and imprecise. When this author compiled a list of radiometric dates for European hominin (both Neanderthals and *H. sapiens*) occupation between c. 65,000 and 20,000 BP (Davies, 2001a), some 2000 determinations were collected from the literature. In the intervening decade, the situation has improved, certainly for Neanderthal sites in western Europe, but coverage is still patchy and concentrated on the period between 60,000 and 25,000 BP. Nevertheless, it would be surprising if there were more than 1000 additional dates, bringing our total to perhaps 3000 determinations. A considerable proportion (at least 35%) of the 2000 dates is unreliable, owing to uncertainties in sample preparation and measurement, often coupled with ambiguous or nonexistent relationships with the archaeology (Pettitt et al., 2003).

The improved potential for the dating of Neanderthal innovations is likely to be restricted to the AMS radiocarbon technique, with its improved sample decontamination (ultrafiltration for bone, antler and ivory, and ABOX for charcoal) and ability to date smaller samples, for the foreseeable future. This means that small samples can be taken from diagnostic organic artefacts (spatulae, spear tips, awls, beads, etc.), and that therefore many of these dates will have much greater archaeological *meaning*. However, it is clear that we cannot rely on AMS radiocarbon alone, and we should also adopt other dating techniques that can allow us to date human innovations directly, such as thermoluminescence dating of burnt lithic artefacts. It is self-evident that our current dating methods do not cover the whole temporal or spatial range of Neanderthals, owing to the inability to obtain precise or accurate ages before 50,000–55,000 BP and to poor collagen preservation in many regions. Changes in our expectations of what we should expect to obtain from dating techniques can, and must, precede improvements in current dating techniques and the development of new methods. In other words, we need to refocus our dating methodologies on grasping the spatio-temporal patterning of hominin behavioural change and development, rather than simply obtaining one or two "range-finding" age estimates for each archaeological "assemblage." The integrity and validity of archaeological assemblages need to be tested and demonstrated, not simply assumed and reinforced; otherwise, we are at best elaborating associations first asserted in the nineteenth and early twentieth centuries, when dating methods were even more rudimentary.

Current techniques allow us, for certain areas and the last millennia of Neanderthal existence, to plot behavioural change over time and space through the direct dating of (organic) artefacts. In some respects, we are also able, through direct dating, to test the validity of our typological and technological categories, and begin to reassess their chronological relationships with each other. We can thus test the validity of (parts of) our presumed assemblages. Given enough dates, we can hope to model the appearance of particular artefact forms and where they arose. Nevertheless, we still need a framework of how Neanderthal and *H. sapiens* societies might have worked in order to interpret our chronologies of innovation. Such frameworks and models are prerequisites for the application of chronology to our study of innovations. Thus, *consistency of approach* can be applied to innovation and inventiveness in the Palaeolithic.

8.5. NEANDERTHAL INNOVATIONS

Much has been made of Neanderthal behavioural innovations in the archaeological record, normally by generalisation rather than focussing on local contingency. Innovations seem to be used as indicators of capacity, rather than as those responses to social and physical environmental demands and possibilities. Such innovations (e.g., Vandermeersch and Maureille, 2007) include:

- Burial of the dead;
- The use of different parts of sites for different functions;
- Blade (and bladelet) technology;
- The use of bone, antler and ivory artefacts and tools;
- The possible use and manufacture of "symbolic" artefacts, e.g., beads;
- Efficient, specialised hunting strategies.

However, few, if any, of the above attributes co-occur in assemblages, and they are discontinuously *distributed* in time and space, probably indicating degrees of cultural variation and different social preferences between Neanderthal groups in various times and places. Our lack of precision in dating, particularly for sites pre-50,000 BP, certainly contributes to such perceptions of spatio-temporal discontinuities in Neanderthal behaviour, leading us to make generalised, direct connections between the intrasite and regional scales, without really evaluating them properly. We take the maximal distribution of Neanderthals in Eurasia, and then assert consistency of behavioural capacities (and, by extension, a sort of universal behaviour) by aggregating behavioural characteristics into trait lists (e.g., Davies and Underdown, 2006; Mellars, 1996; Kozłowski, 1990). The innovations listed above cannot be said to be continually developed; they seem to have been used intermittently, and it is difficult to demonstrate transmission from a source to surrounding populations. Realisation of a capacity is of course dependent on stimuli, whether internally or externally generated. Lamarckian explanations of "inner feeling" cannot be demonstrated for the development of new behaviours, so neo-Darwinian approaches (internally generated novelty, which is selected for by social and/or environmental factors and then transmitted to other social actors) will be discussed here to explain both internal and external generation of behavioural variability.

The list of behaviours outlined above will now be considered in a little more depth. Burial of the Neanderthal dead has been contested in recent years, with some (Gargett, 1989, 1999) arguing that there was no intentional interment in Neanderthal society. However, as has been pointed out (summarised in Pettitt, 2011), burial is practised for a number of reasons, and there is much evidence that Upper Palaeolithic *H. sapiens* buried corpses only intermittently, e.g. the famous ochred burials of the Gravettian (c. 33,000–24,000 BP) (Pettitt, 2011). Neanderthal inhumations are spatially and temporally restricted: most derive from SW France, the Ardennes, the Crimea, the Levant and the Zagros Mountains, and with the possible exception of Tabūn C1 at perhaps 120,000 BP, the majority are dated between 70,000 and 40,000 BP (Maureille, 2011; Maureille and Tillier, 2008; Serangeli and Bolus, 2008; Vandermeersch, 2007; Bar-Yosef and Callander, 1999). In other areas, and sometimes within the same regions, alternative treatments of corpses were practised; the cut marks on Neanderthal bones from SW France, the Rhône valley, Croatia and Spain, for example, suggest that inhumation was not the only way of disposing of the dead. The skull from Kebara 2 is missing, despite the rest of the adjoining bones (including the mandible) remaining undisturbed, but this loss is now thought to have been caused by natural taphonomic processes rather than Neanderthal reworking of the burial (Hovers, pers. comm. in Pettitt, 2011: 115, 118). Other burials might be the result of secondary depositions, to judge from the restricted spatial extents and selective representation of body parts, e.g., Saint-Césaire, La Quina H18, Mezmaiskaya 2, Pech-de-l'Azé 1, Petit-Puymoyen and Sclayn (Maureille, 2011). Burial practices seem to show no clear diachronic development, say from primary (undisturbed) to secondary (reworked) inhumation, and thence to dismemberment/cut marking/cannibalism in any given region. Where such traditions of treating the dead existed, Neanderthals may have drawn on a repertoire of different possibilities, perhaps varied by (emotional) connections with the deceased. It is currently impossible to document the transmission of different characteristics from specific sources to adjoining areas; much more precise dating is needed here to validate or falsify any hypotheses of transmission vs choice from a range of *ad hominem* treatments.

The use of different parts of the site for different functions (structured intrasite variability) also needs much better dating for effective testing in Neanderthal assemblages. It has been described for sites in Portugal (Vilas Ruivas), Spain (Abric Romaní, Roca dels Bous), France (La Folie, Abri de la Combette, Vinneuf, Pucheuil, Bettencourt-St.-Ouen, Villiers-Adam, Éttouteville, Grotte XVI, Beauvais, Abri de Grainfollet, etc.), Ukraine (Molodova I and V) and Israel (Kebara) (Jaubert and Delagnes, 2007). Binford (1998: 234) was sceptical that intrasite spatial organisation was found in non-*H. sapiens* assemblages, using it instead as a defining trait of "cultural modernity". Binford (1998: 229) also points out how hearths and combustion features are major referential features when comparing horizontal structuring of space between sites; archaeologists tend to see them as foci in the microenvironments of sites. Kolen (1999) has coined the term "centrifugal living structures" to describe the tendency of Neanderthals to clear living surfaces by sweeping littering material and waste away from centres of activity, "from the inside to the outside" (p. 148). Often this clearance, rearrangement and reuse of material moved it away from centres (e.g., hearths) towards the peripheries of activities, as part of a continuous, fluid process. Neanderthals, according to Kolen, did not organise their living space "architecturally" (with defined and assigned activity areas) but instead displayed a less structured, flexible and "organic" use of space. This constant reworking of living space, rather than generating clear, classifiable activity patterns, might encourage us to apply neo-Darwinian approaches rather than Lamarckian ones, but diachronic change in Neanderthal site organisation is difficult to identify. Kolen believes that these centrifugal living structures are most convincingly attributed to the Weichselian (certainly with none predating c. 180,000 BP), ending with the structures found in the Châtelperronian (e.g., Grotte du Renne, layers X–IX, at about 45,000–40,000 BP).

The concept of centrifugal living structures has been disputed by other authors (e.g., Conard, 2005), but no evidence has yet come to light that incontrovertibly contradicts the picture of Neanderthal living space as flexibly arranged and reworked. Evidence for intrasite spatial organisation has been discussed for the sites of Abric Romaní (e.g., Carbonell et al., 1996; Rosell et al., 2011; Vaquero et al., 2011) and Tor Faraj (Henry, 1998; Henry et al., 2004), but the findings do not seriously undermine Kolen's hypothesis. Level Ja of Abric Romaní (c. 51,000 BP) seems to show more intrasite structuring of activity areas (around hearths) than the overlying level H (c. 48,000 BP), implying changes in site use and sizes of occupying groups at different periods (Rosell et al., 2011). Floors I and II of Tor Faraj (49,000–69,000 BP) show similar patterning to that seen at the Abric Romaní, with discrete activity areas, centred round hearths, being evident, as well as discard areas towards the margins of the site (Henry, 1998; Henry et al., 2004). In any case, it is probably wisest to treat such centrifugal living structures as proxy evidence for changes in social and economic organisation, rather than representing innovation or inventiveness in their own right. Indirectly, they provide the social locations for innovation, but do not represent clear and widespread novel behaviour themselves. Their characteristics would also imply that there were no *specific* locations for transmission of novel ideas and techniques in Neanderthal sites, though perhaps such transmission might be concentrated in the central areas of greatest activity (of all types).

Blade and bladelet technologies are often thought to be quintessentially characteristic of *H. sapiens*, but an increasing number of such assemblages attributable to Neanderthals have come to light in recent decades (e.g., Révillion and Tuffreau, 1994; Révillion, 1995; Conard et al., 1995; Delagnes et al., 2007; Maíllo Fernández et al., 2004; Slimak, 1999, 2007, 2008; Pastoors and Tafelmaier, 2010). Blades are here taken to be lithic blanks that are at least twice as long as they are wide, and whose production methods create ridges on their dorsal surfaces that run parallel with their margins (Bar-Yosef and Kuhn, 1999). As seen for other traits, there is little diachronic/directional evolution or geographical patterning in the distribution of these Neanderthal blade assemblages, and they are frequently short-lived in particular regions, indicating that their adaptive and developmental potentials were not always self-evident to their manufacturers (Fig. 8.2). Blade technology appears, for example, in northern Europe (the French sites of Séclin, Riencourt-lès-Bapaume, etc.; the German site of Wallertheim) between c. 100,000 and 80,000 BP, but then disappears and is replaced by more "traditional" flake-based Neanderthal assemblages. Blade and bladelet technologies occur elsewhere before this *floruit*, and also afterwards, for example in the Neronian of southern France (perhaps 50,000–45,000 BP) and in the French Châtelperronian (c. 45,000–37,000 BP) (e.g., Slimak, 2007, 2008; Pelegrin, 1995). Should we expect similar stimuli to operate for each reinvention, or were the causes of creation different in every case, perhaps related to the functions served by the technology? Blade production in the Châtelperronian ends with that industry, and is replaced by the Aurignacian (probably *H. sapiens*), while the Neronian at Grotte Mandrin is replaced by five levels of rather more "traditional", flake-based Mousterian before the onset of the Aurignacian (Slimak, 2007). The latter change could have been driven by reduction in mobility and increased territoriality (Slimak, 2007, 2008), leading to greater reliance on more localised materials in the terminal Mousterian at the site, but it could also reflect localised population extinction/displacement: post-Neronian Mousterian industries are only found at one site (Mandrin), and the chronology of the terminal Middle Palaeolithic in the area needs a lot more research.

Blade production cannot be seen as an attempt to economise on raw material, as efficiency can also be gained in flake production (e.g., Eren et al., 2008). Nevertheless, it is worth noting that many blades appear unmodified/unretouched in Neanderthal contexts, despite their margins showing irregularity, which would require regularisation through retouch. Neanderthal blades are often thick as well as irregular in cross section, and laminar production methods very varied (Delagnes et al., 2007). Four, overlapping, methods of Neanderthal blade debitage have been identified, which demonstrate the flexibility and fluidity of their knapping strategies, sometimes oscillating between different ones on the same core (Delagnes et al., 2007), and implying that innovation at smaller (element and morphological) scales of variation was more likely. While knapping strategies in many Middle Palaeolithic blade assemblages might thus lack some clarity/distinctiveness of form and diachronic change – creating extra problems for Lamarckian perspectives on change! – some assemblages do show "Upper Palaeolithic-style" blade production and tool forms (end-scrapers, burins), e.g., Séclin, Riencourt-lès-Bapaume, Wallertheim (Révillion, 1995; Conard et al., 1995). These assemblages are succeeded by more "traditional" Middle Palaeolithic ones, with the laminar industries appearing to disappear quickly. Later on and further south, in SW France, Delagnes et al. (2007) have identified an earlier Middle Palaeolithic phase with a restricted range of (elaborate) debitage and core preparation methods, and with weakly retouched tools, followed by a more recent phase (c. 70–40,000 BP) characterised by a greater diversity of (less-systematic) production modes and increased modification (retouch) of tools. Such modification might show greater evidence of utilisation, maintenance and recycling of toolkits, and might reflect the emergence of new forms of mobility, and a greater flexibility and adaptability to immediate and varied needs during a period of climatic instability. Diversification could be seen as a socially driven innovation, occurring independently of climatic change in the period between 60,000 and 40,000 BP, with selection only becoming more intense as climate began to deteriorate. However, a clearer picture of climatic change in the region for this period needs to be established before we can eliminate climatic factors for element and suite changes.

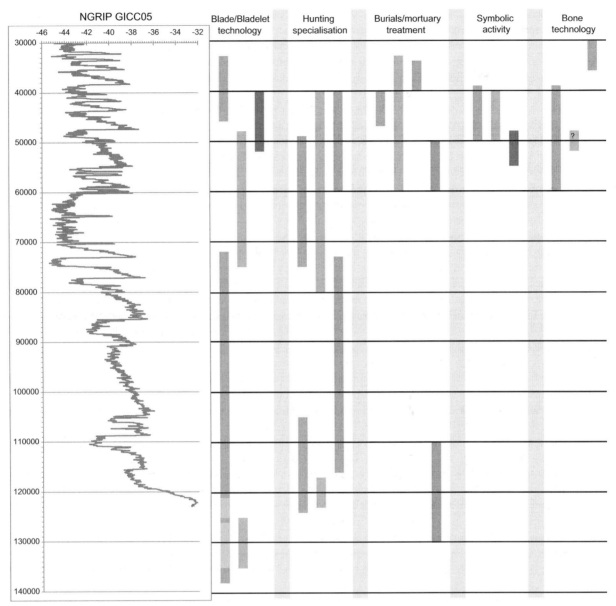

FIGURE 8.2 Spatio-temporal variation in late Neanderthal behavioural innovations, compared against climate (NGRIP members, 2004). Key: Blue = North European Plain (including Ardennes); gold = South-West Europe (southern France, including the Dordogne; Iberia and Italy); brown = South-East Europe (Balkans, Hungary); Mauve = Russian Plain, Crimea; green = Near and Middle East. Semi-shaded areas: inferred persistence of a characteristic in the study area, based on uncertainties attached to the absolute dating, and small sample sizes. This diagram should be interpreted loosely, and is intended only to give an understanding of the *current* chronology of Neanderthal innovations; loosely contemporary traits in the same regions are seldom found in the same assemblages. Date-ranges are largely based upon available absolute age determinations from key sites, and duration of these ranges is often dependent on levels of dating precision and accuracy: for example, the Near Eastern burial period between 110 and 130 ka BP is entirely based on the chronological uncertainty of the Tabūn female Neanderthal remains, while the apparently late Mezmaiskaya child burial in eastern Europe is almost certainly dated too young (Pettitt, 2011). The development of symbolic activity (beads, pendants, colourants, etc.) should not be interpreted literally from this figure, as the number of assemblages with reliable dates is very small; thus, an apparent east-to-west movement of symbolic activity is almost certainly an artefact of our current chronological information (Crémades et al., 1995; d'Errico et al., 2003; Soressi & d'Errico, 2007; Zilhão et al., 2010; Peresani et al., 2011). Fuller references for Neanderthal blade and bladelet technologies, including their chronology, can be found in the main text. "Bone technology" is here defined primarily as bone awls and other shaped (primarily pointed) artefacts (Allsworth-Jones, 1990: 193; d'Errico et al., 1998; Gaudzinski, 1999; Villa and d'Errico, 2001); it must be noted that the SE European bone tools attributed to ~50 ka BP belong to the Jankovichian, and are not directly dated, and dates associated with the bone tools from the Crimean Middle Palaeolithic are likely to be underestimates. "Hunting specialisation" is focussed on assemblages with clear chronological positions, and economic specialisation in particular species and often individuals of specific age and sex (Gaudzinski, 1995, 1996, 2006, etc.).

Bladelets, certainly, are not simply small blades, but have different morphological possibilities (twisted vs curved/straight) and potential functions (e.g., set in multiple, composite arrangements in a wooden haft). The extent of their use in Neanderthal assemblages is debatable, but small lithic blanks seem to have been produced in a small number of known sites (Maíllo Fernández et al., 2004; Maíllo Fernández, 2005; Pastoors and Tafelmaier, 2010). Multiple small sharp edges set into a single shaft allows great flexibility in arrangement and orientation of barbs, allowing in turn greater flexibility of prey species using a restricted range of lithic blank types, but if this insight appeared to Neanderthals, they appear to have made irregular use of it. While Middle Palaeolithic blades were frequently retouched (though cf. Delagnes et al., 2007: 218), bladelets in the same assemblages were much less likely to have been modified, with use-wear studies indicating that many were used in unmodified form (e.g., Slimak, 1999, 2007, 2008; Maíllo Fernández et al., 2004; Pastoors and Tafelmaier, 2010); thus, bladelets are less useful to us for identifying *morphological* change. Neanderthal innovation (at the element and morphological scales), dexterity and technical precision are indicated by the presence of some Neronian bladelets less than 1 cm in length, known as "nanopoints" (Slimak, 2007).

The use of osseous (bone, antler and ivory) materials for the manufacture of tools and artefacts, including symbolic items such as beads and pendants, and also the modification of natural objects (e.g., marine shells) into items of personal adornment, is held to be one of the greatest advances of *H. sapiens* (e.g., Mellars, 1989: 343). The debate has become rather more heated since the late 1990s (d'Errico et al., 1998), when substantial claims for independent invention of such behaviour by Neanderthals have been asserted. Some osseous artefacts appear to lack enough surviving collagen for successful AMS radiocarbon dating (perhaps too old, or owing to poor preservation conditions), but the ones that have been dated directly suggest spatio-temporally varying patterns of use of particular artefact types; the use of beads and pendants is not ubiquitous in Neanderthal contexts, and seems to have been restricted to certain regions (e.g., northern France) and to the end of Neanderthal existence (between c. 45,000 and 36,500 BP) (Higham et al., 2010).

Specialised hunting techniques, focussing on a very narrow range of (herbivore) species, such as reindeer and bison, have been asserted for some Neanderthal assemblages, e.g., Wallertheim, Salzgitter-Lebenstedt and Mauran (e.g., Gaudzinski, 1995, 1996, 2006; Gaudzinski and Roebroeks, 2000; Farizy et al., 1994; Marean, 2005). The term "specialised hunting" needs careful definition; the highly selective species, age and sex profiles of prey could have been partly caused by selective transport of prime carcasses from a broader range of killed individuals elsewhere in the landscape. Nevertheless, some sites do show an in situ butchery of prime individuals (e.g., Wallertheim, horizon B1; late Eemian/early Weichselian), implying hunting selection at or near the kill site (Gaudzinski, 1995). As with other attributes described here, there is no evidence of cumulative selection of prime animals over Neanderthal existence, with examples being spatio-temporally discontinuous. It is thus difficult to gauge the extent and scale of innovation/reinvention in hunting specialisation: how long-lived were such behaviours, and why did they die out? Such subsistence behaviours need to be linked to technological change. Shifts in subsistence between generalisation and specialisation will affect both economy and technology, and we should be able to predict that more specialised technologies will show different, more focused, pathways of innovation (e.g. Straus, 1990), while more generalised technologies will, conversely, show more varied and diffuse ones.

The above traits depend greatly on chronological accuracy and precision, but even given the quality of the dates we have, it is clear that they are intermittently present in the record, rather than consistently developed, disseminated and advanced after their initial appearances. Spatio-temporal gaps in our coverage of Neanderthals, while undoubtedly present, can only provide a very partial explanation of the patterning outlined above; *behavioural characteristics move in and out of the record in the sites already known to us*. Why? Were such fluctuations caused by changing social and/or economic priorities, or to have occurred for reasons of social connectivity or transformations in mobility (e.g., Slimak, 2007, 2008)? Were Neanderthals drawing on *repertoires* of established behavioural traits, with their relative proportions being selected by social and environmental conditions? Local population extinctions, combined with a possible lack of long-distance social networks to transmit information, must also have played a large part in the chequered fortunes of particular traits across time and space, especially given the low population levels generally attributed to Neanderthals. Using mitochondrial DNA, recent estimates of Neanderthal populations have ranged between 3000 and perhaps 12,000 breeding females in the latter part of Neanderthal existence (c. 70,000–35,000 BP) (Noonan et al., 2006; Green et al., 2006, 2008; Briggs et al., 2009). However, these low-density populations were not evenly spread throughout the whole Neanderthal range, and would have been more geographically clustered, perhaps explaining the mtDNA subgroups observed by some researchers: twelve, according to Fabre et al. (2009), covering western Eurasian Neanderthal remains from between 100,000 and 35,000 BP (cf. Briggs et al., 2009). These patchy, localised, populations, often operating at low levels and densities, and with associated risks of localised extinction, represent the social context for innovations we see in Neanderthal assemblages; current evidence would seem to imply that Neanderthal cultural viruses did not outlast their host populations, and thus perished synchronously with their groups.

Similarly evanescent and recurrent patterns of innovation might also be expected for *H. sapiens* populations prior to c. 50,000 BP (e.g., d'Errico et al., 2009). Only with the transition towards more intensive occupations (e.g., the "Developed Aurignacian": Davies, 2001b) in the period between 50,000 and 35,000 BP (depending on location) did *H. sapiens* ensure the more consistent transmission (and survival) of innovations. The mosaic patterning of hominin innovations in the pre-50,000 BP period suggests fluctuating population levels, densities and connectedness, with extinction or contraction-based isolation of populations leading to nontransmission of ideas, or even their extinction if groups found themselves in the wrong place at the wrong time in a period of transition and could not track habitats. By 50,000 BP the range of Neanderthals is contracting, with population displacement or extinction (or both) occurring in Asia, and an apparent "retreat" to Europe; by 40,000 BP, their distribution within Europe has become increasingly patchy. Ironically, the apparent contraction of Neanderthals after c. 50,000 BP into (parts of) Europe might have increased population densities in those areas, and thus facilitated the spread of ideas (see below) in those last groups. After 50,000 BP, both Neanderthals and *H. sapiens* show increasing evidence for symbolic activity, e.g., bead production, though localised in parts of France, in the case of Neanderthals (d'Errico et al., 1998). This paper argues that such increases in symbolic activity can be attributed to increased social interaction, perhaps related to increasing stress on Neanderthal social systems. Such stresses could derive from a combination of climatic and environmental change, coupled with the increasing presence of *H. sapiens* in the landscape at c. 40,000 BP (Stringer and Grün, 1991; Stringer et al., 2003). However, we urgently need much more information on the "density" of such symbolic behaviours over time and space, which requires that more of these behaviours are directly dated, if possible. The dating of the materials from the Grotte du Renne (Higham et al., 2010) marks an important step in this direction, but more such work is needed if we are to gain a clearer chronological framework for shifts in Neanderthal behaviour.

The above discussion might be seen as teleological, in that working generalisations have been adopted (based on a variety of evidence) for the potential social structures and mobilities of Neanderthals. Such generalisations have been employed to make the application of ideas discussed above more straightforward. However, it is possible to discuss innovations without resorting to generalised attributions to particular hominin species; we can treat the evidence for variability in material culture and the potential for social interactions on a site-by-site basis, whether hominin fossils are present or not.

Recent advances in the study of Middle Palaeolithic landscapes have allowed productive lines of enquiry to emerge in Neanderthal research. These also allow us to consider the physical and social environmental impacts of Neanderthal knowledge exchange at intra- and intersite scales. Development of these ideas in future research is thus key to exploring and testing our concepts of information transmission among Neanderthals. A more detailed knowledge of the structures and patterns of Neanderthal landscape use across their whole spatio-temporal extent will allow us to gauge the socioeconomic distances between sites, and permit consideration of how that spatial and seasonal structure might impact on the intensity of face-to-face contact and exchange. Lines of enquiry that connect together raw material provisioning with knapping stages/*chaînes opératoires* and subsistence have recently begun to proliferate in the literature. All provide useful social and demographic context to studies of the generation of novel traits and their transmission or extinction.

White and Pettitt (2011: 77) have recently developed their concept of "local operational areas," which divide landscapes along pathways of movement, and make viewshed connections between sites (assuming few or no trees in the landscape). For example, the Axe valley in Somerset (UK) seems to represent a late Neanderthal Local Operational Area (c. 50,000–45,000 BP), from Hyaena Den and Rhinoceros Hole at the source of the River Axe, to Uphill, where the river would have joined the Bristol Channel plain, some 20 km westwards. This Axe valley Local Operational Area would perhaps have been connected, via residential relocation of Neanderthal groups, to other possible areas some 100–140 km distant, e.g., the Gower peninsula, Tor Bay (White and Pettitt, 2011). Similar analyses of operational areas have also been conducted by researchers in France (Geneste, 1988, 1989; Raynal et al., 2007; etc.), e.g., on early Neanderthal sites (c. 240,000–100,000 BP), and the combination of information from raw material provenance analyses, *chaînes opératoires*, topographic situation of sites and species hunted, together with carcass-processing methods, has proved remarkably instructive (e.g., Raynal et al., 2007). These methods have allowed the testing of Binford's (1979) embedded procurement in the Neanderthal record, and also the reconstructions of pathways in the landscape, based on the stages of the *chaîne opératoire* represented in particular sites, and the sources of lithic raw material that were exploited. At present, dating of these Neanderthal Operational Areas is rather scanty, and would need to be considerably improved if we were to have confidence of site contemporaneity in the distribution of activities across the landscape, but this situation can be improved; the *potential* for increasingly detailed analyses is already present in the addition of other, nearby, archaeological sites. In any case, studies by Raynal et al. (pers. comm.) have already established clear Neanderthal activity patterning in different parts of the southeastern French study region: on current evidence, it would seem that lithic resources from some areas were simply never exploited, and this would have implications for where technological knowledge was developed and transmitted in the landscape.

Layton et al. (in press) have explored the scope for human agency within cognitive and ecological constraints, and have concluded that there are some similarities between Neanderthals and early *H. sapiens*. They see the process of social fission–fusion as important in both species (as well as in chimpanzees), and thus that the process – if not the scale – of mobility was similar in both Neanderthals and *H. sapiens*. Using the encephalisation analysis (neocortex size estimates) of Aiello and Dunbar (1993: 189) to provide an average group size of 145 for Neanderthals, they have analysed the ethnographic high-latitude hunter–gatherer data to provide a guide to Neanderthal demographic densities. The validity of using modern (mostly Arctic Circle) groups as a guide can of course be questioned on many grounds, of which perhaps the most important to us is that for Neanderthals we are considering mid-latitude environments, with normal seasonal variation and higher levels of insolation than can be seen at the poles; thus, environmental productivity would be higher for Neanderthal environments. Nevertheless, in the absence of much work on environmental productivity done for Neanderthal environments (only a little for the period of the final Neanderthals: van Andel and Davies, 2003), we can treat Layton et al.'s comparisons as yielding minimal and maximal figures that can be tested thoroughly against the archaeological settlement models outlined in the preceding paragraph. Assuming, heuristically, that the spatial scale of social interaction in the western Eurasian Middle Palaeolithic was probably comparable to that seen among modern, high-latitude hunter-gatherers, an average of 1 person per 70 km^2 might be assumed. The ethnographic densities collated by Layton et al. range between about 1 person per 10 km^2 (requiring an area of at least 1450 km^2 for a group of 145 people) and 1 person per 200 km^2. For a community of 145 Neanderthals, resource exploitation diameters might have ranged between 100 km (1 person/ 50 km^2) and 200 km (1 person/200 km^2). Following this methodology, denser Neanderthal populations, perhaps reflecting seasonal aggregations, would reflect presence in richer environments, perhaps on the ecotonal boundaries between different habitats; a resource exploitation diameter of 60 km would produce a population density of 1 person/ 20 km^2. Such population densities do not greatly contradict the effective (breeding) population estimates made by mitochondrial geneticists.

Layton et al. point out that if Berger and Trinkaus (1995) are right in their interpretation of the traumas and injuries visible on Neanderthal skeletal remains (both male and female), most of the group was involved with the collection and processing of meat resources. Thus, there was an emphasis on immediate, whole-group, cooperation in food gathering, with little opportunity for delayed reciprocity (the exchange of stored food and social obligations). Whether true or not (and perhaps we should be wary about generalising too much about Neanderthals), these uniformitarian predictions seem to match the transport distances, and discard patterns, seen for lithic raw materials (Féblot-Augustins, 1997). Layton et al. estimate that the minimum foraging range for Neanderthals might have been a 0–6-km radius (40% of deposition events occurring within 10 km), while the maximum one was 0–23 km (64% of deposition events occurring within 25 km). At the larger social scale, the minimum community radius could have been 50 km (80% deposition of material within this radius), with the maximum community radius being 100 km (96% of all material being deposited within this area).

Layton et al.'s argument that Neanderthals should be expected to transport such materials over distances longer than daily foraging radii, and the extent and intensity of this transportation (including artefact forms and any symbolism), has great impact on our understanding of the potential transmission of their ideas. As discussed earlier, if we are looking at intergroup exchange of materials, artefacts or ideas, then we might expect to see "descent with modification" over increasing time and space from the point of origin, while if we see evidence of transmission of novel traits by group fission (and subsequent fusion with other, perhaps distant, groups), then such material could have spread virtually unmodified between disparate groups, apparently in "leaps." Simply because we have evidence of burial, "egalitarian" distribution of traumatic injury throughout groups, and the sharing of resources among Neanderthals – to varying degrees – it does not follow that groups were stable in composition. As Layton et al. point out, modern hunter-gatherers do not spend their whole lives in one band, and it seems sensible to assume that this situation also applied to Neanderthals. Such recombination of groups has major implications for the speed and extent of spread in the transmission of novel traits. We need to clarify whether we expect a new member in a Neanderthal group to modify his or her behaviour to the prevailing *modus operandi*, or whether they would have been content to maintain their own habitual methods. Both scenarios would have the potential to lead to acculturation, and the rate of adoption would surely reflect the numbers of individuals interested in adopting such "new" traits, and the social value placed upon them. Unfortunately, it is not possible to distinguish between either scenario archaeologically: an extensive programme of refitting to allow the identification of individual knappers and their preferred techniques might help to address this issue, if not demonstrate it categorically.

8.6. NEANDERTHALS, CLIMATIC CHANGE AND INNOVATION

It is self-evident that we need more dating of the Neanderthal world – more dates, and better dating accuracy and precision – but we should always remember that new dates are not enough on their own to link behavioural change to wider (environmental) change. We need to revise our own ideas and concepts of the variation in Neanderthal social organisation and mobility, and be very clear about the implications of such variation for the development and dissemination of novel behavioural traits. Generalisations are not particularly useful to us in this regard, and we should be prepared to admit much more fine-scale spatio-temporal variation in the appearance, spread and disappearance of traits. Concepts of inexorable progress should be abandoned, and alternative (neo-Darwinian) selectionist views of generation of variation and varying levels of its transmission and survival be considered instead.

Consideration of nonlinear and nondirectional evolution will allow us to examine changing rates of innovation and processes of demographic change (isolation, social connectivity, extinction and expansion) across the full Neanderthal spatio-temporal range. If we can free Neanderthals from our imposed tyranny of inexorable self-improvement, then we are better equipped to consider their achievements individually within a broader context of social and environmental variability. Currently, it is easier to examine such spatio-temporal variation in the period after 50,000 BP, perhaps because the period is easier to date, though that should not deter us from examining earlier periods of Neanderthal creativity and behavioural flexibility in more detail: are later periods unique, or can they shed light on the less temporally detailed behavioural changes in earlier periods? While, from current data, it might seem as though the intensity of innovation and inventiveness changes and becomes more focussed after about 50,000–45,000 years ago, we cannot be sure at present if this apparent shift is an artefact of our dating, or a behavioural reality. If we can demonstrate more confidence in the latter explanation, we then need to address what might have caused such a shift. One strong possibility would be increased social demand/selection for such change after this time, perhaps related to shifts in mobility, population structure and interactions in the landscape.

At the end of Neanderthal existence, the thorny question of acculturation of European Neanderthals by incoming *H. sapiens* enters the discussion (e.g., d'Errico et al., 1998; Mellars, 1999). The term "acculturation" is used in this chapter simply to denote the transmission of ideas and techniques from members of one group to another ("enculturation" might be seen as occurring within a group). It can occur in a single direction, or operate bidirectionally, depending on the perceived value of the information being transmitted and received. At present, we cannot really say if Neanderthals and *H. sapiens* influenced each other's innovations, or whether they developed independently. On a site-by-site basis, assemblages attributed to *H. sapiens* overlie those of Neanderthals, but absolute dating reveals that, at broader spatial scales, both species occurred in the same or adjoining regions. Indeed, if Neanderthals had become extinct before contact with *H. sapiens*, it would not be possible for exchange of ideas to occur as proposed (d'Errico et al., 1998; Zilhão and d'Errico, 1999). We should not necessarily assume that acculturation can occur simply by picking up the artefacts of an absent (extinct) group, perhaps something akin to Tostevin's (2007) "stimulus diffusion" (Table 8.2); Strathern (1969: 315) describes how a Papua New Guinean lenticular axe found in an area outside that of its production and circulation was not recognised as an axe. Only if we assume that early European *H. sapiens* did not operate clear, "template" views of the tool types they used, might it be possible that they could pick up lithic artefacts made by Neanderthal groups, and replicate them (though would they use them in the same ways?). The well-known example of the Acheulean biface retrieved from elsewhere (the River Vézère?) and deposited in level 2 of the Abri Pataud, after a few (resharpening?) flakes were removed from its tip (Clay, 1995: 83), is useful to our discussion here. There is no evidence that the brief encounter with a pointed biface (familiar to many Neanderthals and earlier hominins) inspired the terminal Gravettian *H. sapiens* of the Last Glacial Maximum to reintroduce such technology to their repertoires. I have touched upon the possibility of inter-Neanderthal acculturation in the preceding sections, but we need a very clear methodology to untangle the evidence for such processes. The archaeological evidence seems to support changes at the *element* and *morphological* scales best, rather than at the suite scale, where clear-cut cases are uncommon, e.g. the Neronian. It is difficult in many situations to distinguish independent (re)invention of characteristics from inter-Neanderthal acculturation.

This chapter has sought to establish that population structure and organisation, rather than simple demic densities or sizes (e.g., Shennan, 2001), were crucial in the transmission and retention of innovations. If we accept that Neanderthals tended to operate face-to-face contact within relatively restricted geographic ranges, with little evidence of transmission along chains of social contacts, then maximum reach of a given innovation will be dependent on the number, age and receptiveness of individuals who come into contact with it. Assuming that a novel trait is recognised

TABLE 8.2 The Impact of Social Connectivity on Degree and Type of Innovation, Linked to Gamble's (1999) social networks (modified from Tostevin, 2007: 345 (Table 28.1))

Degree of Social Intimacy	Social Network Type	Contact Process Resulting From Intimacy	Material Culture Behaviours Affected	Change in Lithic Technology
Low	Global, Extended	Conservatism	The operational sequence and final product remain the same or become entirely novel.	Neither blank production nor toolkit morphology is adopted as a result of little social intimacy or strong cultural resistance.
Low	Extended, Global	Stimulus diffusion	The idea or morphology of the final product is only copied.	Only the toolkit morphology is adopted, leading to equifinal blank production.
Moderate	Effective	Adaptive conservatism	The operational sequence is copied but the morphology of the final product remains unchanged.	Blank production is learnt and replicated but the selected toolkit morphology remains unchanged for reasons of tool function or social signalling.
Medium to high	Intimate, Effective	Diffusion	The operational sequence, the final product, and the social context of use are copied.	Blank production as well as the tool-kit morphology are learnt and replicated.

as such, and then replicated and developed, a balance needs to be struck between intensity of repetition (perhaps very localised) and spatial distribution: innovations that are infrequently used and thinly spread across a large area will be unlikely to survive in much of that range for a long period, though they might be developed and intensified by some users in some places at some times. Conversely, highly localised innovations, which do not spread far, are liable to become extinct, as they stand less chance of developing into extensive cultural viruses. Mobility of individuals and groups is key to maintaining this balance of transmission and development (Table 8.1; Fig. 8.1): group mobility can include both residential (whole-group) and activity-focussed (subgroup) movements, while individual movement is perhaps best exemplified in the form of fission (and subsequent fusion) (Aureli et al., 2008). Mobility would have occurred at different spatial and temporal scales, from daily ranges and encounters of other individuals (low-level transmission of ideas and weak retention of them), through movement and contact extending to weeks or months (medium-level potential for transmission and retention of novel morphological-scale and element-scale changes), to long-term fission–fusion or displacement lasting years or decades that might facilitate the transmission of novel traits at all scales (perhaps even promoting a conceptual shift in the repertoire of the acculturated individual or group).

Although impossible to prove in the Palaeolithic record, we should assume a number of fission–fusion events in an individual's lifetime, probably concentrated in the period of early adulthood. It might be expected that fission–fusion events at a very early age (e.g., adoption) would have little effect on changing the worldview of the individual, as they would simply learn new ideas as if they had been in their original group, and such events would also have little effect on individuals of advanced age, as they would be less inclined to make large-scale changes (though perhaps changes at the morphological and element scales). To make a generalisation useful for further testing and analysis, we might say that we should expect the potential for mobility and fission–fusion, and their impacts upon the generation and transmission of novel traits, to be most pronounced when a Neanderthal individual was aged between perhaps 10–12 and 25 years. Thus, groups with higher levels of such individuals at a given time might be expected to be more experimental or innovative. Obviously, such experimentation has to be balanced against reproductive success, as individuals of this age cohort are also rearing children, and risk taking needs to be managed.

Elaborating on Gamble's (1998, 1999) argument that Neanderthals relied most strongly on intimate and effective networks, based on face-to-face contact, we should expect the bulk of their connections to have been between related (kin) groups. Global networks, as the place where successful innovations really need to reach in order to have a strong chance of long-term survival, would barely exist, if at all, in the Neanderthal world. A social world comprising many small-scale closed networks, with limited exchange of information and ideas between them, cannot be seen as the ideal substrate for the long-term persistence of new ideas and techniques. This structure seems to be the best explanation we have at present for the patterning seen in Neanderthal assemblages. Even the evidence for symbolism in (late) Neanderthal contexts seems to be localised and varied by region, implying that bead use, for example, did not extend much beyond northern and parts of southwestern France (d'Errico et al., 1998). Many other innovations, such as the Neronian, were even more localised. Individuals whose skills and abilities were more trusted by others would presumably have been imitated more, and thus their novel traits were more likely to have been transmitted, though it is very unlikely that any one person would have been skilled in everything. Face-to-face transmission would therefore have been the main process in the transmission of ideas from skilled to less-skilled artisans in Neanderthal groups and networks.

The work of Tostevin (2003, 2007, etc.) has encouraged us to break down innovation and inventiveness in the archaeological record into individual elements, which can be quantified to facilitate intercomparisons between knapping choices and *chaînes opératoires* in different assemblages. Such quantification and description are essential if we are to have confidence in identifying different scales of innovation, and thus address possible social processes behind their adoption and development. Are all suite-scale changes simply the result of demic extinction/displacement and subsequent replacement, or can we envisage a situation where a group would change all key cultural markers? Quantification of such change is now possible (Table 8.2), but answers are not necessarily any easier. For example, if we can eliminate a direct (causal) link between changes in the physical environment and behaviour, it would seem justifiable to attribute shifts in the latter to internal, social, drivers, but that does not mean that "environment" and "society" should be treated as two independent variables. Both were, and are, interlinked. Our perceptions of change in the physical environment are conditioned by social preoccupations, and there is little reason to suppose that Neanderthals were better at such perception than we are.

8.6.1. Was Behavioural Innovation Inevitable?

After novel traits have been generated, selection processes begin to operate, not only on whether the behaviour is replicated, but also on how it is developed and transmitted, *and to whom*. This socioenvironmental selection is unlikely to have initiated innovation, but to have had greater effect on extinction, as Foley (1994) has observed. This two-stage generation and selection process means that we need to distinguish circumstances leading to innovation from those that lead to the retention, transmission and development of those innovations. Different stimuli and processes can explain each stage in innovation: one set of environmental (social or physical) stimuli can help to explain why a particular innovation arises (e.g., raw material parsimony); different social factors might elucidate why it is replicated, adopted and developed; and entirely different reasons determine its spread and ultimate success or failure. Much of this detail will be lost to archaeologists, but the situation is certainly not hopeless; given good chronology and detailed environmental records, it will be possible to consider the role of climatic and environmental factors in the development of an innovation.

Tostevin's (2007) analyses encourage us to consider how differing levels of social intimacy and contact could have affected the method and quality/type of innovation (Table 8.2). Unless social intimacy is medium to high, as in Gamble's intimate networks, adoption of a whole suite of characteristics (large-scale change) is unlikely to occur. At lower levels of social intimacy, perhaps reflecting Gamble's extended and global networks, diffusion of traits, if it happens at all, is more superficial, restricted to the replication of form. Following Gamble's argument, if the structure of larger-scale Neanderthal social networks (i.e., effective and extended ones) were to alter, then archaeological signatures of change would have to adjust accordingly, or even disappear (Fig. 8.1). In other words, behavioural innovation should be seen as occurring within a system of covarying attributes, which can be variously combined, inflected and emphasised with retained elements.

Neo-Darwinian approaches can envisage innovation in both active/"hot" and passive/"cold" (Bender, 1985) decision-making situations; Cullen's cultural virus theory emphasises the role of conscious decisions in the selection and replication of ideas (see above). Our general association of innovation with creativity and replication and development of ideas might lead us to suppose that it should be seen as an active process, practised by "hot" societies in full control of their social destinies. I should, however, like to propose that innovation can also be produced passively (as long as it is retained, replicated and subjected to variation). The key point, in neo-Darwinian terms, is that variability is constantly generated, upon which all types of environmental selection can operate. It is a mistake to assume that diffusing innovations (see Table 8.2) and conservatism must directly indicate active and passive decisions, respectively; in many cases, innovations could be seen as selectively neutral, at least in terms of the physical environment, if not the social one. It is perfectly possible for poor "active" decisions to be made, which must be either adapted or abandoned in the face of adverse (selective) conditions. Innovations are simply part of the series of social negotiations that hominins have to make continuously with their environments, which may or may not lead to identifiable (directional) change. It would seem that the careers of most innovations, like those of their hominin progenitors, "end in failure." The archaeological record does not tell us which innovations were successful and which were not; both can be preserved, and it is for us to infer what falls into which category. Reiteration of innovations can tell us something about the "success" of an innovation, even if it has to be periodically reinvented. However, reiteration and reinvention might also reflect underlying behavioural conservatism – anathema to any researcher inclined to see innovations as "progressive" (i.e., Lamarckian). Behavioural "conservatism" can sometimes support its population perfectly adequately. The contemporary late Neanderthal groupings of Châtelperronian in France and terminal Mousterian industries in southern Iberia might be superficially seen as "progressive" and "conservative" populations, respectively, but in terms of their longevity, the Iberian tortoise outran the French hare. The series of innovations (notably increased blade production and personal adornment) did not necessarily benefit the Châtelperronian French populations in the long term: some of them were certainly widespread, but not long-lived. In order to explain such patterning, we need to return to the localised, rather than to the generalised, levels, even when trying to explain larger (suite) scales of change. Approaches such as those of Raynal et al. (2007) need to be more widely adopted, so that we can talk about local networks and mobility with greater confidence; these can then provide the analytical framework for discussing the iteration and spread (if at all) of innovative behaviours.

For heuristic purposes, it might be easiest to assert that Neanderthal cultural viruses (if they existed) were very short term in their effects. (Re)invention in different times and places should falsify the presence of cultural viruses in Neanderthals, as the latter would require continuity of use and development to sustain them outside their originating host population. Repeated invention of novel traits might thus be documenting successive events of population reorganisation, perhaps including local extinction. In order to evaluate whether a population extinction event may be contemporaneous with extinction of its innovations (e.g., the Neronian), we need a coherent strategy to obtain more data: both an improved chronological framework and a larger number of Neronian sites in the region. An increased spatio-temporal detail for case-study entities such as the Neronian and Châtelperronian would allow us to examine the mechanisms for the appearance and disappearance of novel traits with much greater confidence.

8.7. FUTURE DIRECTIONS

With recent advances in modelling capacity, Hägerstrand-type cellular models, with a temporal dimension, have begun to be applied to generate rates and intensity of spread for particular innovations in the Palaeolithic (e.g., Powell et al., 2009). So far, it has not been possible to test these models with any confidence, as we need an extensive series of directly dated behavioural characteristics, wherever possible, to achieve that goal. In future, these models will need to consider the following characteristics if they are to describe the hunter–gatherer social world in a convincing manner:

- Realistic treatments of hunter–gatherer movement need to be modelled. We require more detailed considerations of how Neanderthals in different spatio-temporal zones might have structured their landscape use. Powell et al. (2009) assume movement between fixed social points in the landscape for up to 100 generations (c. 2000 years!), but this is essentially an agriculturalist view of human mobility. Population displacements need to be modelled, so that shifting human occupation in the landscape in response to changing resource distributions can be used in models of the spread of novel traits. In certain environmental conditions, hominin groups might have been more evenly distributed in the landscape than at other times, when there were unoccupied areas. Such variable presence would have affected proximity of groups, and thus of transmission of cultural information. Thus, these models require greater clarity on what is meant by "migration" (a term that encompasses many spatio-temporal scales), and the intensity and duration of population interactions.
- Zubrow's (1989) modelling of late Neanderthal and early *H. sapiens* interactions in Europe was also based on stable and stationary populations, with no migration. His results indicated that Neanderthal mortality was 1–2% higher than that seen in *H. sapiens*, also adversely affecting their life expectancy. These simulations would need to be rerun with consideration taken of fission–fusion in groups and varying levels of mobility, before we can assess the implications of longer-lived *H. sapiens* for the transmission of information over longer temporal distances. If grandparents are involved in the transmission of ideas, the patterning in the replication, reiteration and spread of novel traits would be entirely different from those groups where longevity was more restricted, with fewer generations being copresent. Zubrow modelled the interaction of two hominin species; it would also be interesting to compare the effects of competition *between* contemporary Neanderthal groups.
- Predictive and generative models of the spread of novel traits also need to consider site sizes, layouts and behavioural signatures more carefully, if we are to evaluate their impact on the transmission of innovations. The numbers of people copresent (especially important in face-to-face transmission), and how they interact physically, will be of crucial importance in the spread of innovations.

- Models should also consider the impact of economic "specialisation" vs "generalisation" in the origination and spread of novel traits. We should not expect successful innovations in "specialist" economies to be the same as those in more "generalist" ones, with novel traits in the former being perhaps more "focussed" and largely restricted to behaviours related to the main economic activities, with those in the latter being more evenly spread. Thus, economic impacts on innovation should be expected to vary in material culture intensity and spatio-temporal spread.

In the current absence of extensive reliable chronologies (especially for all but the last few millennia of Neanderthal existence), we shall need to run many series of predictive and generative models to explore some of the ideas about innovation expressed above. However, we shall ultimately need to test these models against the archaeological record, and that means we also have to rethink our perspectives on, and methods of, measuring variability in the record. There are exciting times ahead of us for exploring Neanderthal creativity, and in seeking to relate it to changing social and physical environments.

NOTE

All radiometric date estimates given in this paper are in calendric years for ease of comparison between radiocarbon and other dating techniques. Radiocarbon dates have been converted to calendar-age estimates using the CalPal 2007 programme (GICC05-Hulu curve: Weninger et al., 2007).

ACKNOWLEDGEMENTS

I wish to thank the editor for his patience in the long gestation of this paper, and the reviewer for many useful suggestions. I also thank those who undertook the arduous task of reading sections, and my students, on whom I tested many of the ideas. All residual errors and omissions are, of course, my responsibility.

REFERENCES

NGRIP [North Greenland Ice Core Project] members, 2004. High-resolution record of northern hemisphere climate extending into the last interglacial period. Nature 431, 147–151.

Aiello, L., Dunbar, R., 1993. Neocortex size, group size, and the evolution of language. Curr. Anthropol. 34, 184–193.

Allsworth-Jones, P., 1986. The Szeletian and the Transition from Middle to Upper Palaeolithic in Central Europe. Oxford University Press, Oxford.

Allsworth-Jones, P., 1990. The szeletian and the stratigraphic succession in Central Europe and adjacent areas: main trends, recent results, and problems for resolution. In: Mellars, P. (Ed.), The Emergence of Modern Humans. Edinburgh University Press, Edinburgh, pp. 161–242.

Anderson, D.G., Gillam, J.C., 2000. Paleoindian colonization of the Americas: implications from an examination of physiography, demography, and artefact distribution. Am. Antiq. 65, 43–66.

Aureli, F., Schaffner, C.M., Boesch, C., Bearder, S.K., Call, J., Chapman, C.A., Connor, R., Di Fiore, A., Dunbar, R.I.M., Henzi, S.P., Holekamp, K., Korstjens, A.H., Layton, R., Lee, P., Lehmann, J., Manson, J.H., Ramos-Fernandez, G., Strier, K.B., van Schaik, C.P., 2008. Fission–fusion dynamics: new research frameworks. Curr. Anthropol. 49, 627–654.

Bar-Yosef, O., Callander, J., 1999. The woman from Tabun: Garrod's doubts in historical perspective. J. Hum. Evol. 37, 879–885.

Bar-Yosef, O., Kuhn, S.L., 1999. The big deal about blades: laminar technologies and human evolution. Am. Anthropol. 101, 322–338.

Bender, B., 1985. Prehistoric developments in the American midcontinent and in Brittany, Northwest France. In: Price, T.D., Brown, J.A. (Eds.), Prehistoric Hunter–Gatherers: The Emergence of Cultural Complexity. Academic Press, Inc, San Diego, pp. 21–57.

Bennett, K.D., Provan, J., 2008. What do we mean by 'refugia'? Quaternary Sci. Rev. 27, 2449–2455.

Berger, T.D., Trinkaus, E., 1995. Patterns of trauma among the Neandertals. J. Archaeol. Sci. 22, 841–852.

Bettinger, R.L., 1991. Hunter–Gatherers: Archaeological and Evolutionary Theory. Plenum Press, New York (Interdisciplinary Contributions to Archaeology).

Binford, L.R., 1979. Organization and formation processes: looking at curated technologies. J. Anthropol. Res. 35, 255–273.

Binford, L.R., 1980. Willow smoke and dogs' tails. Am. Antiq. 45, 4–25.

Binford, L.R., 1998. Hearth and home: the spatial analysis of ethnographically documented rock shelter occupations as a template for distinguishing between human and hominid use of sheltered space. In: Conard, N.J., Wendorf, F. (Eds.), Middle Palaeolithic and Middle Stone Age Settlement Systems (Proceedings of the XIII Congress of the International Union of Prehistoric and Protohistoric Science, Forlí, Italy). A.B.A.C.O. Edizioni, Forlí, pp. 229–239.

Bordes, F., 1961. Mousterian cultures in France. Science 134, 803–810.

Breuil, H., 1912. Les subdivisions du Paléolithique supérieur et leur signification. Congrès International d'Anthropologie et d'Archéologie Préhistoriques [XIVe session, Geneva], I: 165–238.

Briggs, A.W., Good, J.M., Green, R.E., Krause, J., Maricic, T., Stenzel, U., Lalueza-Fox, C., Rudan, P., Brajković, D., Kućan, Ž., Gušić, I., Schmitz, R., Doronichev, V.B., Golovanova, L.V., de la Rasilla, M., Fortea, J., Rosas, A., Pääbo, S., 2009. Targeted retrieval and analysis of five Neandertal mtDNA genomes. Science 325, 318–321.

Bronk Ramsey, C., Higham, T., Bowles, A., Hedges, R., 2004. Improvements to the pretreatment of bone at Oxford. Radiocarbon 46, 155–163.

Carbonell, E., Cebrià, A., Allué, E., Cáceres, I., Castro, Z., Díaz, R., Esteban, M., Ollé, A., Pastó, I., Rodríguez Álvarez, X.P., Rosell, J., Sala, R., Vallverdú, J., Vaquero, M., Vergés, J.M., 1996. Behavioural and organizational complexity in the middle Palaeolithic from Abric Romaní. In: Carbonell, E., Vaquero, M. (Eds.), The Last Neandertals, the First Anatomically Modern Humans. Cultural Change and Human Evolution: The Crisis at 40 ka BP. Igualada, Barcelona, pp. 385–434.

Chauvière, F.-X., 2005. Quand le technique jalonne le temps: la notion de "temps technique" en archéologie paléolithique. Bulletin De La Société Préhistorique Française 102, 757–761.

Clarke, D.L., 1968. Analytical Archaeology. Methuen & Co, London.
Clay, R.B., 1995. Le Protomagdalénien de l'abri Pataud niveau 2. In: Bricker, H.M. (Ed.), Le Paléolithique Supérieur De L'abri Pataud (Dordogne): Les Fouilles De H.L. Movius Jr. Suivi D'un Inventaire Analytique Des Sites Aurignaciens Et Périgordiens De Dordogne. Éditions de la Maison des Sciences de l'Homme, Paris, pp. 67–87 (Documents d'Archéologie Française, 50).
Conard, N.J., 2005. An overview of the patterns of behavioural change in Africa and Eurasia during the middle and late Pleistocene. In: d'Errico, F., Backwell, L. (Eds.), From Tools to Symbols: From Early Hominids to Modern Humans. Witwatersrand University Press, Johannesburg, pp. 294–332.
Conard, N.J., Preuss, J., Langohr, R., Haesaerts, P., van Kolfschoten, T., Becze-Deak, J., Rebholz, A., 1995. New geological research at the middle Palaeolithic locality of Wallertheim in Rheinhessen. Archäologisches Korrespondenzblatt 25, 1–27.
Crémades, M., Laville, H., Sirakov, N., Kozłowski, J.K., 1995. Une pierre gravée de 50 000 ans B.P. dans les Balkans. Paléo 7, 201–209.
Cullen, B.S., 2000. Contagious Ideas: On Evolution, Culture, Archaeology, and Cultural Virus Theory. Oxbow Books, Oxford.
Dalén, L., Nyström, V., Valdiosera, C., Germonpré, M., Sablin, M., Turner, E., Angerbjörn, A., Arsuaga, J.L., Götherström, A., 2007. Ancient DNA reveals lack of postglacial habitat tracking in the arctic fox. Proc. Natl. Acad. Sci. 104, 6726–6729.
Davies, W., 2001a. The Archaeological Database [Stage 3 Project]. http://wserv2.esc.cam.ac.uk/research/research-groups/oistage3/stage-three-project-database-downloads (accessed 30.05.11).
Davies, W., 2001b. A very model of a modern human industry: new perspectives on the origins and spread of the aurignacian in Europe. Proc Prehistoric Soc. 67, 195–217.
Davies, R., Underdown, S., 2006. The Neanderthals: a social synthesis. Camb. Archaeol. J. 16, 145–164.
de Loecker, D., 1994. On the refitting analysis of Site K: a Middle Palaeolithic findspot at Maastricht-Belvédère (The Netherlands). Ethnographisch-Archäologische Zeitschrift 35, 107–117.
Delagnes, A., Jaubert, J., Meignen, L., 2007. Les technocomplexes du Paléolithique moyen en Europe occidentale dans leur cadre diachronique et géographique. In: Vandermeersch, B., Maureille, B. (Eds.), Les Néandertaliens: Biologie et Cultures. Éditions du Comité des travaux historiques et scientifiques, Paris, pp. 213–229 (Documents Préhistoriques 23).
Dyson-Hudson, R., Smith, E., 1978. Human territoriality: an ecological assessment. Am. Anthropol. 80, 21–41.
d'Errico, F., Henshilwood, C., Lawson, G., Vanhaeren, M., Tillier, A.-M., Soressi, M., Bresson, F., Maureille, B., Nowell, A., Lakarra, J., Backwell, L., Julien, M., 2003. Archaeological evidence for the emergence of language, symbolism, and music—an alternative multidisciplinary perspective. J. World Prehistory 17 (1), 1–70.
d'Errico, F., Zilhão, J., Julien, M., Baffier, D., Pélegrin, J., 1998. Neanderthal acculturation in western Europe? A critical review of the evidence and its interpretation. Curr. Anthropol. 39 (supplement), S1–S44.
d'Errico, F., Vanhaeren, M., Barton, N., Bouzouggar, A., Mienis, H., Richter, D., Hublin, J.-J., McPherron, S.P., Lozouet, P., 2009. Additional evidence on the use of personal ornaments in the Middle Paleolithic of North Africa. Proc. Natl. Acad. Sci. U S A 106, 16051–16056.

Eren, M.I., Greenspan, A., Sampson, C.G., 2008. Are upper Paleolithic blade cores more productive than Middle Paleolithic discoidal cores? A replication experiment. J. Hum. Evol. 55, 952–961.
Fabre, V., Condemi, S., Degioanni, A., 2009. Genetic evidence of geographical groups among Neanderthals. PloS One 4 (4), e5151 (doi:10.1371/journal.pone.0005151).
Faivre, J.-P., 2006. L'industrie moustérienne du niveau Ks (locus 1) des Fieux (Miers, Lot): mobilité humaine et diversité des compétences techniques. Bulletin de la Société Préhistorique Française 103, 17–32.
Farizy, C., David, F., Jaubert, J., 1994. Hommes Et Bisons Du Paléolithique Moyen À Mauran (Haute Garonne). CNRS Éditions, Paris.
Féblot-Augustins, J., 1997. La Circulation Des Matières Premières Au Paléolithique. E.R.A.U.L. 75, Liège.
Finlayson, C., 2004. Neanderthals and Modern Humans: An Ecological and Evolutionary Perspective. Cambridge University Press, Cambridge (Cambridge Studies in Biological and Evolutionary Anthropology).
Foley, R.A., 1994. Speciation, extinction, and climatic change in hominid evolution. J. Hum. Evol. 26, 277–289.
Gamble, C., 1998. Palaeolithic society and the release from proximity: a network approach to intimate relations. World Archaeol. 29, 426–449.
Gamble, C., 1999. The Palaeolithic Societies of Europe. Cambridge University Press, Cambridge.
Gamble, C., 2007. Origins and Revolutions: Human Identity in Earliest Prehistory. Cambridge University Press, Cambridge.
Gargett, R.H., 1989. Grave shortcomings: the evidence for Neandertal burial. Curr. Anthropol. 30, 157–190.
Gargett, R.H., 1999. Middle Paleolithic burial is not a dead issue: the view from Qafzeh, Saint-Césaire, Kebara, Amud, and Dediriyeh. J. Hum. Evol. 37, 27–90.
Gaudzinski, S., 1995. Wallertheim revisited: a re-analysis of the fauna from the Middle Palaeolithic site of Wallertheim (Rheinhessen, Germany). J. Archaeol. Sci. 22, 51–66.
Gaudzinski, S., 1996. On bovid assemblages and their consequences for the knowledge of subsistence patterns in the middle Palaeolithic. Proc. Prehistoric Soc. 62, 19–39.
Gaudzinski, S., 1999. Middle Palaeolithic bone tools from the open-air site Salzgitter-Lebenstedt (Germany). J. Archaeol. Sci. 26, 125–141.
Gaudzinski, S., 2006. Monospecific or Species-dominated faunal assemblages during the middle Paleolithic in Europe. In: Hovers, E., Kuhn, S.L. (Eds.), Transitions before the Transition: Evolution and Stability in the Middle Paleolithic and Middle Stone Age. Springer, New York, pp. 137–147 (Interdisciplinary Contributions to Archaeology).
Gaudzinski, S., Roebroeks, W., 2000. Adults only. Reindeer hunting at the middle Palaeolithic site Salzgitter Lebenstedt. J. Hum. Evol. 38, 497–521.
Geneste, J.-M., 1988. Systèmes d'approvisionnement en matières premières au paléolithique moyen et au paléolithique supérieur en aquitaine. In: Kozłowski, J.K. (Ed.), 1988. L'homme De Neandertal: La Mutation, vol. 8. E.R.A.U.L. 35, Liège, pp. 61–70.
Geneste, J.-M., 1989. Economie des ressources lithiques dans le moustérien du sud-ouest de la France. In: Freeman, L.G., Patou, M. (Eds.), 1989. L'homme de Neandertal: La Subsistance, vol. 6. E.R.A.U.L. 33, Liège, pp. 75–97.

Geneste, J.-M., 1991. Systèmes techniques de production lithique: variations techno-économiques dans les processus de réalisation des outillages lithiques. Techniques et Cultures 17/18, 1–35.

Green, R.E., Krause, J., Ptak, S.E., Briggs, A.W., Ronan, M.T., Simons, J.F., Du, L., Egholm, M., Rothberg, J.M., Paunovic, M., Pääbo, S., 2006. Analysis of one million base pairs of Neanderthal DNA. Nature 444, 330–336.

Green, R.E., Malaspinas, A.-S., Krause, J., Briggs, A.W., Johnson, P.L.F., Uhler, C., Meyer, M., Good, J.M., Maricic, T., Stenzel, U., Prüfer, K., Siebauer, M., Burbano, H.A., Ronan, M., Rothberg, J.M., Egholm, M., Rudan, P., Brajković, D., Kućan, Ž., Gušić, I., Wikström, M., Laakkonen, L., Kelso, J., Slatkin, M., Pääbo, S., 2008. A complete Neandertal mitochondrial genome sequence determined by high-throughput sequencing. Cell 134, 416–426.

Hägerstrand, T., 1967a. Innovation Diffusion as a Spatial Process [Translated by A. Pred & G. Haag]. University of Chicago Press, Chicago.

Hägerstrand, T., 1967b. On Monte Carlo Simulation of Diffusion. In: Garrison, W.L., Marble, D.F. (Eds.), Quantitative Geography, Part I: Economic and Cultural Topics. Northwestern University Studies in Geography, number 13, Evanston, IL, pp. 1–32.

Harris, M., 1979. Cultural Materialism: The Struggle for a Science of Culture. Vintage Books, New York.

Henneberg, M., 2009. Two interpretations of human evolution: essentialism and Darwinism. Anthropol. Rev. 72, 66–80.

Henry, D.O., 1998. Intrasite spatial patterns and behavioral modernity: indications from the late levantine mousterian rockshelter of Tor Faraj, Southern Jordan. In: Akazawa, T., Aoki, K., Bar-Yosef, O. (Eds.), Neandertals and Modern Humans in Western Asia. Plenum Press, New York, pp. 127–142.

Henry, D.O., Hietala, H.J., Rosen, A., Demidenko, Y.E., Usik, V.I., Armagan, T.L., 2004. Human behavioral organization in the middle Paleolithic: were Neanderthals different? Am. Anthropol. 106, 17–31.

Higham, T., 2011. European middle and Upper Palaeolithic radiocarbon dates are often older than they look: problems with previous dates and some remedies. Antiquity 85, 235–249.

Higham, T., Jacobi, R., Julien, M., David, F., Basell, L., Wood, R., Davies, W., Bronk Ramsey, C., 2010. Chronology of the grotte du renne (France) and implications for the context of ornaments and human remains within the châtelperronian. Proc. Natl. Acad. Sci. U S A 107, 20234–20239.

Isaac, G., 1981. Stone age visiting cards: approaches to the study of early land use patterns. In: Hodder, I., Isaac, G., Hammond, N. (Eds.), Pattern of the Past: Studies in Honour of David Clarke. Cambridge University Press, Cambridge, pp. 131–155.

Jaubert, J., Delagnes, A., 2007. De l'espace parcouru à l'espace habité au paléolithique moyen. In: Vandermeersch, B., Maureille, B. (Eds.), Les Néandertaliens: Biologie et Cultures. Éditions du Comité des travaux historiques et scientifiques, Paris, pp. 263–281 (Documents Préhistoriques 23).

Jochim, M.A., 1983. Palaeolithic cave art in ecological perspective. In: Bailey, G. (Ed.), Hunter-gatherer Economy in Prehistory: A European Perspective. Cambridge University Press, Cambridge, pp. 212–219.

Jochim, M., 1987. Late Pleistocene refugia in Europe. In: Soffer, O. (Ed.), The Pleistocene Old World: Regional Perspectives. Plenum Press, New York, pp. 317–331 (Interdisciplinary Contributions to Archaeology).

Kolen, J., 1999. Hominids without homes: on the nature of middle Palaeolithic settlement in Europe. In: Roebroeks, W., Gamble, C. (Eds.), The Middle Palaeolithic Occupation of Europe. University of Leiden, Leiden, pp. 139–175.

Kozłowski, J.K., 1990. A multiaspectual approach to the origins of the Upper Palaeolithic in Europe. In: Mellars, P. (Ed.), The Emergence of Modern Humans: An Archaeological Perspective. Edinburgh University Press, Edinburgh, pp. 419–437.

Kuhn, S.L., 2003. In what sense is the levantine initial upper Paleolithic a "transitional" industry? In: Zilhão, J., d'Errico, F. (Eds.), The Chronology of the Aurignacian and of the Transitional Technocomplexes: Dating, Stratigraphies, Cultural Implications. Instituto Português de Arqueologia, Lisboa, pp. 61–69 (Trabalhos de Arqueologia 33).

Lamarck, J.B.P.A., 1809. Philosophie Zoologique, Tome Premier. Dentu, Paris.

Lartet, É., Christy, H., 1864. Sur des figures d'animaux gravées ou sculptées et autres produits d'art et d'industrie rapportables aux temps primordiaux de la période humaine [published separately as "Cavernes du Périgord: objets gravés et sculptés des temps préhistoriques dans l'Europe occidentale"]. Révue Archéologique 9, 233–267.

Lartet, É., Christy, H., (ed. T.R. Jones)1875 [1865–1875]. Reliquiæ Aquitanicæ; being contributions to the archæology and palæontology of the Périgord and the adjoining provinces of southern France. London: Williams & Norgate.

Layton, R., O'Hara, S., Bilsborough, A. Antiquity and social functions of multi-level social organisation among human hunter-gatherers. In: Grueter, C.C., van Schaik, C.P., Zinner, D. (Eds.), International Journal of Primatology: special issue on multi-level societies, in press.

Leroi-Gourhan, A., 1964. Le Geste Et La Parole: Technique Et Langage. Éditions Albin Michel, Paris.

Lieberman, D.E., Shea, J.J., 1994. Behavioral differences between archaic and modern humans in the levantine mousterian. Am. Anthropol. 96, 300–332.

Lubbock, J., 1864. Cave-men. Nat. Hist. Rev. 4, 407–428.

Maíllo Fernández, J.M., 2005. Esquemas operativos líticos del Musteriense final de Cueva Morín (Villanueva de Villaescusa, Cantabria). Museo de Altamira, Monografías 20, 301–313.

Maíllo Fernández, J.M., Cabrera Valdès, V., Bernaldo de Quirós, F., 2004. Le débitage lamellaire dans le moustérien final de cantabrie (Espagne): le cas de El Castillo et Cueva Morín. L'Anthropologie 108, 367–393.

Marean, C.W., 2005. From the tropics to the colder climates: contrasting faunal exploitation adaptations of modern humans and Neanderthals. In: d'Errico, F., Backwell, L. (Eds.), From Tools to Symbols: From Early Hominids to Modern Humans. Witwatersrand University Press, Johannesburg, pp. 333–371.

Maureille, B., 2011. Les sépultures des néandertaliens et autres gestes envers les morts. Dossiers D'Archéologie (Néandertal Réhabilité) 345 (May/June), 40–47.

Maureille, B., Tillier, A.-M., 2008. Répartition géographique et chronologique des sépultures nénadertaliennes. In: Vandermeersch, B., Cleyet-Merle, J.-J., Jaubert, J., Maureille, B., Turq, A. (Eds.), Première Humanité: Gestes funéraires des Néandertaliens. Éditions de la Réunion des musées nationaux, Paris, pp. 66–74 (Musée National de Préhistoire, Les Eyzies-de-Tayac).

Mayr, E., 1970. Populations, Species, and Evolution. An Abridgment of Animal Species and Evolution. Belknap Press (Harvard University Press), Cambridge.

Mellars, P., 1989. Technological changes across the Middle-Upper Palaeolithic transition: economic, social and cognitive perspectives. In: Mellars, P., Stringer, C. (Eds.), The Human Revolution: Behavioural and Biological Perspectives in the Origins of Modern Humans. Edinburgh University Press, Edinburgh, pp. 338–365.

Mellars, P., 1996. The Neanderthal Legacy: An Archaeological Perspective from Western Europe. Princeton University Press, Princeton (NJ).

Mellars, P., 1999. The Neanderthal problem continued [with comments]. Curr. Anthropol. 40, 341–364.

Noonan, J.P., Coop, G., Kudaravalli, S., Smith, D., Krause, J., Alessi, J., Chen, F., Platt, D., Pääbo, S., Pritchard, J.K., Rubin, E.M., 2006. Sequencing and analysis of Neanderthal genomic DNA. Science 314, 1113–1118.

Nowell, A., White, M., 2010. Growing up in the middle Pleistocene: life history strategies and their relationship to Acheulian industries. In: Nowell, A., Davidson, I. (Eds.), Stone Tools and the Evolution of Human Cognition. University of Colorado Press, Boulder, pp. 67–82.

Pastoors, A., Tafelmaier, Y., 2010. Bladelet production, core reduction strategies, and efficiency of core configuration at the middle Palaeolithic site Balver Höhle (North Rhine Westphalia, Germany). Quartär 57, 25–41.

Peresani, M., Fiore, I., Gala, M., Romandini, M., Tagliacozzo, A., 2011. Late Neandertals and the intentional removal of feathers as evidenced from bird bone taphonomy at Fumane Cave 44 ky B.P., Italy. Proc. Natl. Acad. Sci. U S A 108, 3888–3893.

Pettitt, P., 2011. The Palaeolithic Origins of Human Burial. Routledge, Abingdon.

Pettitt, P.B., Davies, W., Gamble, C.S., Richards, M.B., 2003. Palaeolithic radiocarbon chronology: quantifying our confidence beyond two half-lives. J. Archaeol. Sci. 30, 1685–1693.

Powell, A., Shennan, S., Thomas, M.G., 2009. Late Pleistocene demography and the appearance of modern human behavior. Science 324, 1298–1301.

Raichlen, D.A., Armstrong, H., Lieberman, D.E., 2011. Calcaneus length determines running economy: implications for endurance running performance in modern humans and Neandertals. J. Hum. Evol. 60, 299–308.

Raynal, J.-P., Fernandes, P., Santagata, C., Guadelli, J.-L., Moncel, M.-H., Patou-Mathis, M., Fernandez, P., Fiore, I., 2007. Espace mineral et espace de subsistence au paléolithique moyen dans le sud du massif central français: les sites de sainte-anne I (haute loire) et de payre (ardèche). In: Moncel, M.-H., Moigne, A.-M., Arzarello, M., Peretto, C. (Eds.), Aires D'approvisionnement en Matières Premières Et Aires D'approvisionnement en Ressources Alimentaires: Approche intégrée des comportements, pp. 141–159. Oxford: B.A.R. International Series 1725.

Révillion, S., 1995. Technologie du débitage laminaire au paléolithique moyen en Europe septentrionale: état de la question. Bulletin de la Société Préhistorique Française 92, 425–441.

Révillion, S., Tuffreau, A., 1994. Les Industries Luminaires au Paléolithique Moyen. CNRS Éditions, Paris (Dossier de Documentation Archéologique no. 18).

Roebroeks, W., Kolen, J., van Poecke, M., van Gijn, A., 1997. 'Site J': an early weichselian (Middle Palaeolithic) flint scatter at maastricht-belvédère, the Netherlands. Paléo 9, 143–172.

Rosell, J., Blasco, R., Fernández-Laso, M.C., Vaquero, M., Carbonell, E., 2011. Connecting areas: faunal refits as a diagnostic element to identify synchronicity in the Abric Romaní archaeological assemblages. Quaternary Int. http://dx.doi.org/10.1016/j.quaint.2011.02.019.

Rowley, S., 1985. Population movements in the Canadian Arctic. Études Inuit Studies 9 (1), 3–21.

Semenov, S.A., 1964. Prehistoric Technology: An Experimental Study of the Oldest Tools and Artefacts from Traces of Manufacture and Wear [Translated from Pervobytnaiatekhnika (1957) by M.W. Thompson]. Cory, Adams & Mackay, London.

Serangeli, J., Bolus, M., 2008. Out of Europe – the dispersal of a successful European hominin form. Quartär 55, 83–98.

Shennan, S., 2000. Population, culture history, and the dynamics of culture change. Curr. Anthropol. 41, 811–835.

Shennan, S., 2001. Demography and cultural innovation: a model and its implications for the emergence of modern human culture. Camb. Archaeol. J. 11, 5–16.

Slimak, L., 1999. Mise en evidence d'une composante laminaire et lamellaire dans un complexe moustérien du Sud de la France. Paléo 11, 89–109.

Slimak, L., 2007. Le Néronien et la structure historique du basculement du Paléolithique moyen au Paléolithique supérieur en France méditerranéenne. Comptes-Rendus Palevol 6, 301–309.

Slimak, L., 2008. The neronian and the historical structure of cultural shifts from middle to Upper Palaeolithic in Mediterranean France. J. Archaeol. Sci. 35, 2204–2214.

Soressi, M., d'Errico, F., 2007. Pigments, gravures, parures: les comportements symboliques controversies des néandertaliens. In: Vandermeersch, B., Maureille, B. (Eds.), Les Néandertaliens: Biologie et Cultures. Éditions du Comité des travaux historiques et scientifiques, Paris, pp. 297–309 (Documents Préhistoriques 23).

Spuhler, J.N., 1984. *The evolution of apes and humans: genes, molecules, chromosomes, anatomy and behaviour [The Douglas Ormonde Butler lecture for 1984]*. Brisbane.

Strathern, M., 1969. Stone axes and flake tools: evaluations from New Guinea. Proc. Prehistoric Soc. 35, 311–329.

Straus, L.G., 1990. The original arms race: Iberian perspectives on the Solutrean phenomenon. In: Kozłowski, J.K. (Ed.), Feuilles de Pierre: Les Industries à Pointes Foliacées du Paléolithique Supérieur Européen, pp. 425–447. Liège: E.R.A.U.L. 42.

Stringer, C., Grün, R., 1991. Time for the last Neanderthals. Nature 351, 701–702.

Stringer, C., Pälike, H., van Andel, T.H., Huntley, B., Valdes, P., Allen, J.R.M., 2003. Climatic stress and the extinction of the Neanderthals. In: van Andel, T.H., Davies, W. (Eds.), Neanderthals and Modern Humans in the European Landscape during the Last Glaciation: Archaeological Results of the Stage 3 Project. McDonald Institute for Archaeological Research Monographs, Cambridge, pp. 233–240.

Tostevin, G.B., 2003. A Quest for antecedents: a comparison of the terminal middle Palaeolithic and early Upper Palaeolithic of the levant. In: Goring-Morris, A.N., Belfer-Cohen, A. (Eds.), More than Meets the Eye: Studies on Upper Palaeolithic Diversity in the Near East. Oxbow Monographs, Oxford, pp. 54–67.

Tostevin, G.B., 2007. Social intimacy, artefact visibility and acculturation models of neanderthal-modern human interaction. In: Mellars, P., Boyle, K., Bar-Yosef, O., Stringer, C. (Eds.), Rethinking the Human Revolution: New Behavioural and Biological Perspectives on the Origin and Dispersal of Modern Humans. McDonald Institute for Archaeological Research Monographs, Cambridge, pp. 341–357.

Vandermeersch, B., 2007. Les néandertaliens du proche-orient et de l'asie du sud-ouest. In: Vandermeersch, B., Maureille, B. (Eds.), Les Néandertaliens: Biologie et Cultures. Éditions du Comité des travaux historiques et scientifiques, Paris, pp. 87–94 (Documents Préhistoriques 23).

Vandermeersch, B., Maureille, B. (Eds.), 2007. Les Néandertaliens: Biologie et Cultures. Éditions du Comité des travaux historiques et scientifiques, Paris (Documents Préhistoriques 23).

Vanhaeren, M., d'Errico, F., 2006. Aurignacian ethno-linguistic geography of Europe revealed by personal ornaments. J. Archaeol. Sci. 33, 1105–1128.

Vaquero, M., Chacón, M.G., García-Antón, M.D., Gomez de Soler, B., Martínez, K., Cuartero, F., 2011. Time and space in the formation of lithic assemblages: the example of Abric Romaní level J. Quaterary Int. http://dx.doi.org/10.1016/j.quaint.2011.12.015.

Verpoorte, A., 2007. Neanderthal energetics and spatial behaviour. Before Farming 2006/3, Article 2 http://www.waspress.co.uk/journals/beforefarming/journal_20063/abstracts/index.php (accessed 30.05.11).

Villa, P., d'Errico, F., 2001. Bone and ivory points in the Lower and Middle Paleolithic of Europe. J. Hum. Evol. 41, 69–112.

Weninger, B., Jöris, O., Danzeglocke, U., 2007. CalPal-2007 (Cologne Radiocarbon Calibration & Palaeoclimate Research Package). http://www.calpal.de/ (accessed 30.05.11).

White, M.J., Pettitt, P.B., 2011. The British Late Middle Palaeolithic: an interpretative synthesis of Neanderthal occupation at the northwestern edge of the Pleistocene world. J. World Prehistory 24, 25–97.

Zilhão, J., d'Errico, F., 1999. The chronology and taphonomy of the earliest Aurignacian and its implications for the understanding of Neandertal extinction. J. World Prehistory 13, 1–68.

Zilhão, J., 2006. Aurignacian, behavior, modern: issues of definition in the emergence of the European upper Paleolithic. In: Bar-Yosef, O., Zilhão, J. (Eds.), Towards a Definition of the Aurignacian: Proceedings of the Symposium Held in Lisbon, Portugal, June 25–30, 2002. Trabalhos de Arqueologia 45, Lisboa, pp. 53–69.

Zilhão, J., Angelucci, D.E., Badal-García, E., d'Errico, F., Daniel, F., Dayet, L., Douka, K., Higham, T.F.G., Martínez-Sánchez, M.J., Montes-Bernárdez, R., Murcia-Mascarós, S., Pérez-Sirvent, C., Roldán-García, C., Vanhaeren, M., Villaverde, V., Wood, R., Zapata, J., 2010. Symbolic use of marine shells and mineral pigments by Iberian Neandertals. Proc. Natl. Acad. Sci. U S A 107, 1023–1028.

Zubrow, E., 1989. The demographic modelling of Neanderthal extinction. In: Mellars, P., Stringer, C. (Eds.), The Human Revolution: Behavioural and Biological Perspectives in the Origins of Modern Humans. Edinburgh University Press, Edinburgh, pp. 212–231.

Index

Note: Page numbers with "f" denote figures; "t" tables.

A

Acculturation, of Neanderthals, 10, 39–40, 44, 106–107, 119–120
Alternative reality, archeology of, 96–98
Anatomically modern humans (AMH), 3–4, 6–9, 15, 18, 20, 23–24, 26, 28–31, 36–39, 41, 43–46, 51–52, 54, 61, 89–93, 95–98
Ape toolmaking, 58
Artefacts, 1, 3–4, 6–11, 94
Aterian industry, 10–11, 23–34, 25
    environmental context for developments in, 20
    innovations in, 17, 19–20, 26, 56, 61, 70–71, 93, 110, 113
    revised chronology for, 18–19
Aurignacian, 8, 18, 30, 40–41, 43–45, 79
Australopithecus aethiopicus, 5, 54
Australopithecus garhi, 5, 54
Australopithecus, 3, 5, 54
Axe valley Local Operational Area, 118–119

B

Behavioural change and diversity, 1, 23, 56, 72, 74–76, 103–113, 120
Behavioural conservatism, 122–123
Behavioural innovation, inevitability of, 122–123
Behavioural modernity, 3–4, 7–8, 17, 23, 24, 36–37, 39, 104
Bifacial tools, 11, 24–26, 31, 39, 41, 72, 75, 77, 95
Birch bark pitch, 8, 38
Blade production, of Neanderthals, 8, 10, 24–25, 38, 113, 115–118
Blade production, Upper Paleolithic, 17–18, 44, 115–116
Blades and bladelet technology, 8, 10, 17–18, 24–25, 38, 44, 113, 115–118
Body painting, Aterian culture, 23
Body painting, of Neanderthals, 9, 42–43
Brain
    physical evolution of, 1–5, 2f, 16, 18–20, 55, 80, 90–94, 96–98
    in primates, 1–4, 80, 91
    social, 3, 18–19, 94, *see also* Minds, hominins
Burials, of Neanderthals, 8, 10, 28, 37, 37f, 39–41, 113–114, 116, 119

C

Canyon type of space, 53
Centrifugal living structures, 114–115
Chaîne opératoire, 105–106, 118–119
Châtelperronian industry, 8–9, 30, 37–38, 38f, 42f, 39–46
Climate
    creativity and competition, 9, 103–128
        behavioural innovation, inevitability of, 122–123
        dating, precision and accuracy in, 113
        future directions, 123–124
        Neanderthal innovations, 113–122
        population expansion and contraction, social processes of, 11, 30–31, 82, 107–112, 115
        possible causes and scales of change, 103–107
    effect on Neanderthals, 9–10, 113–122
Codons, 89
Cognition
    Cognition, creative, 52
    demographics and, 7–8, 80
    distributed, 16
    fluidity, 15
Cognition, Neanderthal, 9, 39, 43
Cognitive evolution, 1, 3, 61, 72–74, 80
Collective consciousness, 3, 16
Colourants, 9, 26–28, 27f, 28f, 38, 42, 46, 116f
Competition, effect on Neanderthals, 9–10, 103, 123
Composite implements, 4, 17, 95, 97–98, 105, 117
Connectivity, effects of, 17, 76–80, 112, 120–121
Cooperative foraging, 4
Core tool, 6, 10–11, 24–25, 55–56, 58, 59f, 61, 75, 95, 110, 115
Courtship displays, 90–92, 95, 98
Creative act, 4–5, 18, 51–54, 62, 70, 72
Creative explosions, 6, 69, 71
Creative novelty, 51–52, 56
Creativity, 1, 3–11, 51–54
    aggregate, 70–71, 82
    definitions of, 15, 69, 89
    as emergent phenomena, 72–74, 95–96
    emergent, 72, 81
    evidence for, in western Europe, 97
    exceptional, 59, 61–62
    evolutionary ecology of, *see* Evolutionary ecology, of creativity
    hominin, 16
    mundane creativity, 53–54, 61
    Neanderthal, 112, 120, 124
    Oldowan lithic technology, 52–54
    realised creativity, 51–52
    symbolic, 17
Cultural behaviors, 16–17
Cultural evolution, 6, 16, 51, 53–54, 71–72, 74, 81–82
Cultural virus theory, 112, 121–123

D

Dating, precision and accuracy in, 7–10, 24, 26, 40, 43–45, 106, 113–114, 117–118, 120
Demic expansion, 6–7, 9, 30–31, 74, 79–82, 103–104, 112
Demographics, 6–7, 9, 15–18, 30–31, 74, 79–82, 103, 112
    and human cognition, 7
    and novel traits, 9
    and symbolic thought, 16–17
    and technological innovation, 6–7
Dietary niche, broadening of, 56–57
Dual-inheritance theory, 16, 73–74

E

Embedded procurement, 109, 118
Emotions, and social bonds, 3, 15–16, 19–20, 19t, 112, 114
Environment, 4–6
    and developments in Aterian industry, 18, 20, 29–30
    response to, 4–9, 11, 56, 71, 74–82, 91–94, 103–105, 110, 112–114, 117–124
Environment, cultural, 45

Essentialism, 106
Eukaryotes, 90
Evo-demo, 74
Evolution to order, 104–105
Evolutionary ecology, of creativity, 4, 89–102
    alternative reality, archeology of, 96–98
    foraging strategy and information sharing, 92–94, 98
    hierarchical organization, 3, 19, 89–98
    modern humans as "major transition" in evolution, 90–91
    pair bonding and courtship displays, 91–98
    phenotypic thought, emergence of, 94–96

F

Face-to-face communication, 9, 20, 73, 77, 80, 109, 111–112, 118, 120–123
Fire technology, 6, 7–8, 38, 70–71
Foraging strategy, 3–4, 56–58, 61, 76–77, 81, 90–94, 98, 119
Frontal lobes, 1, 96

G

Genetic information, 1, 91
Genetics, 6, 9, 16, 31, 36, 38–39, 46, 72–74, 80, 89–93, 98, 98, 107, 119
Group living, 3, 6, 15, 18–19, 60, 63, 76–78, 80, 93, 103–113, 115, 118–123
Group mobility, 120–121

H

Hägerstrand's model, 109, 111, 123
Hand axes, 4, 94–95, 97
Hierarchies
    of emotions, 19t
    in personal networks, 19
Homing space, 54, 59, 61
Hominin mobility, scales of, 110t, 111t
Hominin-made lithic assemblages, 54
Hominins, 2–3, 5, 7, 17–20, 23, 52, 54–62, 70–72, 75–77, 79, 94, 103, 106, 112, 120, 122
Homo sp., 5, 54–55, 98
Homo erectus, 1–3, 5, 19, 54, 58, 61, 77
Homo ergaster/erectus, 76–77
Homo sapiens, *see also* Anatomically Modern Humans, 1–2, 24, 26, 120
    Neanderthal society and, similarities between, 119
    transmission of innovations, 118, 120
Honeybees
    information-centre foraging, 4, 91–93
Humans
    language, 1–2, 4, 18, 36, 89–98
        versus waggle dance, 93
    modern, as "major transition" in evolution, 90–91
    societies, *see also* Societies, 4
Humpback whale, transformation of neuronal information in, 91, 97
Hunter–gatherer populations, 4, 7, 75–78, 94, 103, 109, 119, 123
Hylobates, 91–92

I

Ideas transmission, different scales of hominin mobility on, 110t
Impoverishment
    of earliest Lower Palaeolithic artefact, 75
    of East Asian early Palaeolithic technologies, 75
Information
    sharing, 92–94, *see also* Neuronal information

Information center-foraging, 92–93
Information overload, 97–98
Innovation, 4–6, 11
Innovation, cultural, 61
    definition of, 15, 104
    degree and type of, 121t
    as emergent phenomenon, 71–74
    versus inventiveness, 105
    large-scale change, 105
    medium-scale change, 105
Innovation, Neanderthal, 104–107, 108f, 109, 113–118, 120, 123
    in mobile groups, 107–112
    rates, variation in, 74, 76, 79–81
    small-scale change, 105
Interaction, *see* Social interaction
Invention, 3–10, 19, 38, 51–53, 57, 61–62, 69, 72, 76, 80–82, 104, 112, 117, 120, 123
Inventiveness versus innovation, 105

K

Klondike spaces, 53–54

L

Lamarckism
    on cultural issues, 106–107
    and innovations, 104–105, 114–115, 123
Language, 1–2, 4, 7, 18, 36, 89–98
Large-scale social networks, 78, 80
Leapfrog/saltation model, 109
Levallois industry, 10, 24–26, 71–72, 75

M

McElreath's solution, 80
Meat consumption, 5–6, 56–58, 62, 77, 119
Megaptera novaeangliae, 91
Mental template, 4, 91, 94–95
Middle Palaeolithic, 6, 9–10, 17–18, 23–24, 37–39, 41–44, 46, 70–71, 74–75, 115–116, 118–119
Middle Palaeolithic toolkits, 10, 71, 74–75, 78
Middle Stone Age, North Africa during, 6–7, 10, 16–18, 23, 39, 71, 74, 78
Minds, hominins, 3–4, 11, 15–20, 112
    distributed, 16
    modelling, 15–16
    number of, 16–18
Mitochondrial Eve hypothesis, 35, 38–39
Mobile groups, innovation in, 107–112
Mobility
    of individuals/groups, 60, 76–78, 104, 107–112, 115, 117, 119–121
    and novel traits, 9
Modern behaviour, *see* Modernity
Modern hunter-gatherers, 103, 119–120
Modernity, 7–8, 18, 30–31, 36, 79, 96
Modernity, anatomical, 39–40
Modus operandi, 119–120
Mousterian Industry, 9, 24–26, 28, 30–31, 37, 39, 41–44, 46, 71–72, 115, 123
Mundane creativity, 5–6, 53–54, 61–62
Music, 17, 20, 69, 89, 91–92, 96–97

N

Nanopoints, 117
Nassarius shells, 1, 11, 17, 27f, 28, 30–31, 39, 45–46

Index

Neanderthal
  acculturation, 120
  behavioural innovation, inevitability of, 122–123
  blades and bladelet technology, 115
  climatic change and innovation, 104, 120–123
  culture, 6, 71, 74–75, 79, 105
  dating, precision and accuracy in, 113
Neanderthal, extinction of, 103–104
  habitat distributions, tracking, 103–104
  Homo sapiens and, similarities between, 119
  innovations, 113–120
    spatio-temporal variation in, 116f
Neanderthal, interbreeding with AMH, 9
  population expansion and contraction, social processes of, 107–112
    innovation in mobile groups, 107–112
    possible causes and scales of change, 104–107
Neanderthal, society, 103–128
  specialised hunting techniques, 117
Neanderthals, 6, 8f
  behaviour, 37–39
  effect of climate and competition on, 9–10
  personal ornaments and symbolism among, 7, 10, 31, 35–49
Nebular spatial distributions, 109
Networking, human, 4, 7, 90, 94
Networks, effects of, 76–80
Neuronal information, 90–92
  sharing, 92–94
  transformation of, 91
Noise
  Lamarckian view, 106, 107, 110
  neo-Darwinian view, 107
Non-Homo stone toolmaker, 54–55
North Africa
  during Middle Stone Age, 16–18, 75
  origins of symbolically mediated behaviour, 10–11, 23–34, 97

O

Oasis type of space, 53
"Older than the Oldowan" hypotheses, 57
Oldowan lithic technology, 51–62, 75, 77
  creativity, 52–54
  dietary niche, broadening of, 56–57
  Oldowan, overview of, 54–56
  recognisable archaeological entity, 52
  re-invention and cultural transmission, 59–61
  spatial clusters, occurrence of, 56
  systematic stone toolmaking, 57–59
  technocomplex, 59–60
  tools, 4–6, 94–95
    assemblages, validity of, 5
    raw materials used for, 5
    stone tools, 59
"Ordinary creative use of language,", 91

P

Pair bonding, 91–92
Palaeoanthropology, 3, 7–9, 11, 51, 69–72
Paranthropus boisei, 5, 54, 57
Parietal cortex, 1–2
Passive modification, 58
Pedunculate tools, 11, 24–25, 25f, 26–28, 31
Personal networks, active, 18–19

Personal ornaments
  in Aterian industry, 23, 28, 31
  among Neanderthals, 7–9, 35–49
Phenotypic thought, emergence of, 94–96
Plateau type of space, 53
Population genetics, forces of, 73–74
Population size and robustness, on behavioural change and diversity, 7, 17, 60, 74–76, 78, 81–82
Population stability, 75
Possibilities space
  modelling of, 53
  topography of, 54
Post-Howieson's Poort, 71
Prepared-core techniques, 95
Primates, 1–5, 18–20, 23, 36, 52, 56–60, 62, 80, 90–93
  brain evolution, 3

R

Radical novelty, 53, 59
Realised creativity, 51–52
Recent African Origin (RAO) model, 35–37
Reciprocal altruism, 94
Refuge
  active models of, 112
  concept of, 103–104
Re-invention (reinvention), 4–6, 31, 52, 59, 61–62, 104, 115, 117, 123

S

Senses, 19–20
Small-world networks, 77, 79
Small-world topologies, 77–78
Social behaviour, 3–4, 19, 60
Social brains, 3, 18–19, 94
Social complexity, 3–4, 15, 19–20
Social interaction, 3, 6–7, 10, 19–20, 46, 60, 76, 80–81, 118–119
Social intimacy, 121t, 122
Social learning, 7, 16–17, 52, 60, 72, 80
Social networks, 4, 7, 31, 74, 76–80, 90, 94, 108f, 111, 117, 121t
Social networks, Neanderthal, 9, 121t, 122
Societies, 4, 6, 16, 23, 40, 73, 76–78, 81, 90, 93, 104, 107, 109, 113, 122
  Neanderthal, 10, 46
  Primate, 4, 90
Spatial clusters, occurrence of, 56
Specialised hunting techniques, 117
Stability of transmission, 112
Stone tool technology, *see also* Aurignacian Industry, Chatelperronian Industry, Mousterian Industry, 1–6, 10, 93
  Neanderthals, and symbolic material culture, 40–51, 46
  Oldowan, 4–6, 52, 54–62
  prepared-core techniques, 95
  problem of stasis in, 3–4, 60, 62, 76
String of pearls model, 108f, 109
Symbolically mediated behaviour, North African origins of, 10–11, 23–34
Symbolism, 96
  and humans, 10, 23–24, 28
  among Neanderthals, 7–10, 35–49, 43f, 44f, 119, 122
Syntactic language, 89, 93, 96–98
Systematic stone toolmaking, 57–59

T

Technologies, early, 6–11
Theory of mind (ToM), 19t, 20, 54–55, 62, 93–94, 98

Tool making, 2–4, *see also* Stone tool technology
Tool making, by non-human primates, 52
Tortoise core, 10–11
Transmission, cultural, 5, 17, 30–31, 52, 59–62, 75–76, 79, 106
Transmission of ideas, mobility and potential for, 7, 10, 17, 57, 74, 108f, 109–111, 114–115, 118, 120–123
Transmission of knowledge, 6, 54
Transmission of traits, 9, 16, 112
Transmission systems, long-distance mobility effects on, 112

U
Upper Palaeolithic, 4, 6–8, 19, 30, 36, 41–42, 44f, 69–71, 74, 78–79, 114
Upper Palaeolithic (UP) transition, 17, 17t, 40, 78–79, 106

W
Waggle dance of bees versus human language, 93
Weichselian, 103, 114, 117
Wilderness type of space, 53